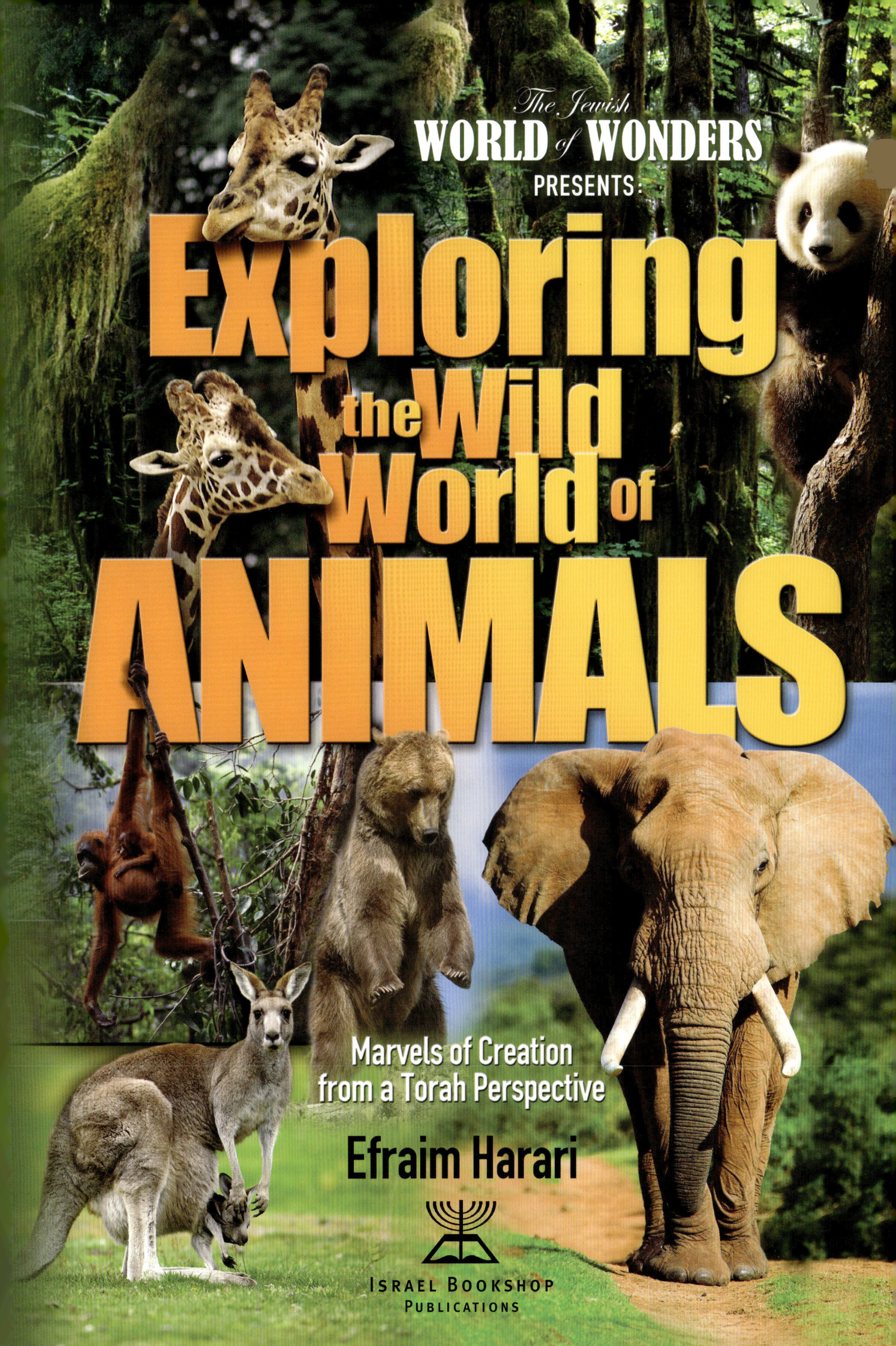

Copyright © 2012 by Israel Bookshop Publications

ISBN 978-1-60091-233-7

All rights reserved. No part of this book may be reproduced or transmitted in any form or by any means (electronic, photocopying, recording or otherwise) without prior permission of the publisher.

Compiled and adapted by Efraim Harari
Layout and Design by Stacey Gindi,
www.SGindiDesignStudio.com

Torah sources verified by Rabbi Michael Malka
Edited by Malkie Gendelman
Cover Design by Sruly Perl · 845-694-7186

Published by:
Israel Bookshop Publications
501 Prospect Street
Lakewood, NJ 08701

Tel: (732) 901-3009
Fax: (732) 901-4012
www.israelbookshoppublications.com
info@israelbookshoppublications.com

Printed in China

Distributed in Israel by:
Shanky's
Petach Tikva 16
Jerusalem
972-2-538-6936

Distributed in Europe by:
Lehmanns
Unit E Viking Industrial Park
Rolling Mill Road,
Jarrow, Tyne & Wear NE32 3DP
44-191-430-0333

Distributed in Australia by:
Gold's Book and Gift Company
3- 13 William Street
Balaclava 3183
613-9527-8775

Distributed in South Africa by:
Kollel Bookshop
Northfield Centre
17 Northfield Avenue
Glenhazel 2192
27-11-440-6679

Dedication

Since this is my first book, I was not quite sure to whom I should dedicate the book.

I did some research and found that most writers chose one of the following people to dedicate their books to: a) a family member, b) the person who helped them create the book, c) their best friend, d) their business partner, e) their role model, or f) the person who inspires them most. For most people, this decision is an extremely difficult one.

For me, *baruch Hashem*, the choice was obvious.

This book is dedicated in honor of my best friend, my life partner, my role model, and the person I most admire – my devoted wife:

Chava bat Esther

A true *eshet chayil*

May the *zechuyot* emanating from the study of this book be a source of blessing and success in all her endeavors...

Table of

Preface 6

 **Apes 7**

 **Elephant 49**

 **Bears 19**

 **Fox 55**

 **Beaver 31**

 **Gazelle 61**

 **Camel 37**

 **Giraffe 67**

 **Cow 43**

 **Goat 73**

Contents

 **Hippopotamus** 79

 **Monkeys** 115

 **Horses** 85

 **Porcupine** 127

 **Kangaroo** 97

 **Snakes** 133

 **Leopard** 103

 **Weasel** 145

 **Lion** 109

 **Wolf** 151

Sources 157

Photo Credits 158

Preface

"Hashem made the beast of the earth according to its kind, and the animal according to its kind, and every creeping being of the ground according to its kind, and Hashem saw that it was good."

(Bereishis 1:25)

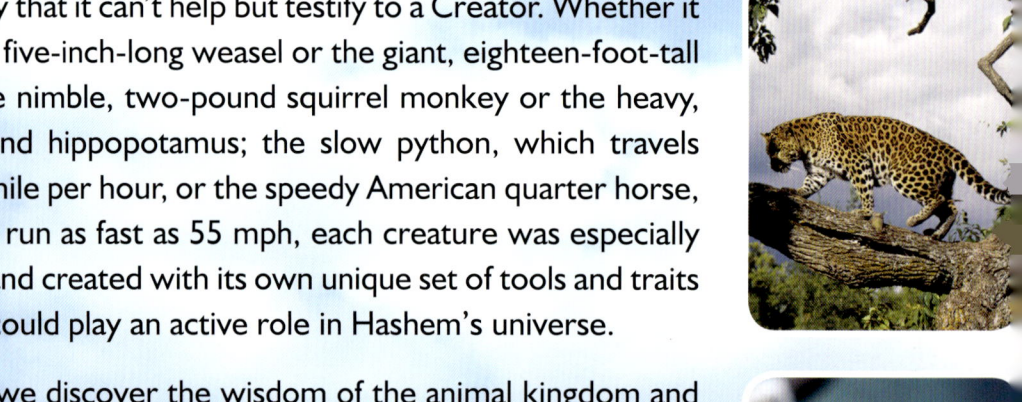

The animal kingdom is full of such great variety, design, and complexity that it can't help but testify to a Creator. Whether it is the tiny, five-inch-long weasel or the giant, eighteen-foot-tall giraffe; the nimble, two-pound squirrel monkey or the heavy, 7,000-pound hippopotamus; the slow python, which travels only one mile per hour, or the speedy American quarter horse, which can run as fast as 55 mph, each creature was especially designed and created with its own unique set of tools and traits so that it could play an active role in Hashem's universe.

Join us as we discover the wisdom of the animal kingdom and marvel at such diverse traits as the teeth of the beaver, the trunk of an elephant, the quills of a porcupine, the pouch of a kangaroo, the mane of the lion, the udder of a cow, the hump of the camel, as well as the slyness of the fox, the boldness of the leopard, the grace of the gazelle, the beauty of the Appaloosa horse, and the intelligence of the monkey and the ape. Each animal is so unique, so different, yet all share the same fundamental function: to reveal the greatness and omnipotence of Hashem.

Inside this book, we will take you on a spectacular journey around the globe, as we travel through jungles, forests, mountains, deserts, and the open fields to explore the wonders and design of thirty-five different members of the animal kingdom. Besides enjoying the incredible photos, you will learn fascinating facts and interesting insights into Hashem's amazing creatures.

APE קוף

Ape / קוף

AMAZING! Chimps use medicinal plants to treat themselves for illness and injury.

Fast Fact: Chimpanzees can be taught human sign languages such as ASL (American Sign Language).

The Chimpanzee

Apes Versus Monkeys

Apes and monkeys are primates, which means that they possess characteristics such as five-fingered hands with opposing thumbs, forward-facing eyes, and color vision. Apes, however, are not monkeys! Monkeys are more likely to be in trees and use their tails for balance. Apes do not have tails, and are much larger than monkeys, with the exception of gibbons. But the most important difference is that apes are more intelligent than monkeys. Their brains are larger and more developed. The apes include gorillas, chimpanzees, bonobos, orangutans, and gibbons.

The Chimpanzee

Chimpanzees, commonly referred to as chimps, have long arms and short legs. Their faces are pinkish to black, and their bodies are covered with long black hair. Their opposable toes and thumbs help them hold objects easily. Chimps are quadrupedal, which means they walk on all four limbs, although they are able to walk upright for short distances. Male chimps are about four feet tall and weigh between 90 and 120 pounds.

Chimps live in dense rainforests, woodlands, swamps, and grasslands, and can be found in twenty-one African countries. They live in community groups called *troops* that have up to eighty members. These large groups consist of smaller family groups of three to six chimps.

During the day, chimps spend most of their time in trees; however, they travel mostly on the ground. Chimps are agile climbers and build their nests high up in trees as well as on the ground. They like to eat fruit, leaves, nuts, seeds, ants, and termites. Sometimes they supplement their diet with meat from goats, antelopes, or other primates.

Chimps are very intelligent and are able to use various sounds and gestures to communicate with other chimps. They also use many different objects as tools, much more so than any other animal. They'll use grass stems or twigs as "fishing rods," poking them into termite or ant nests and eating the insects that cling to them. They are able to use a rock as a hammer and break open nuts that are encased in tough shells. If enemies approach, they will defend themselves by throwing sticks and stones at them.

Did You Know?
Each chimpanzee has its own individual pant-hoot, a common call used in different situations.

No Cry Zone: Gorillas do not shed tears when they cry.

Baby Fact: Young gorillas, from three to five years old, spend most of their day at play. Their daily schedule consists of climbing trees, chasing each other, and swinging on branches.

The Gorilla

Gorillas are the largest of all primates. A male gorilla can be as tall as six feet and weigh up to six hundred pounds. Adult female gorillas are about half the size of the males. Gorillas usually walk on all fours, but frequently stand upright. Most of their bodies are covered by thick, dark hair. An adult male becomes a "silverback" between the ages of twelve and fifteen, when he is fully grown and the hair on his back turns silvery-gray.

Not all gorillas look the same. Differences include length and color of hair, jaw size, facial shape, and arm length. Due to these differences, zoologists have placed gorillas into three subspecies. They are the western lowland gorilla, the eastern lowland gorilla, and the mountain gorilla. Their names indicate where they live in Africa.

Gorillas can climb trees, but are usually found on the ground in communities of up to thirty members. These troops are very well organized and are led by one dominant silverback. The silverback makes all the decisions, such as where the troop will travel for food each day, when to stop to eat or rest, and where to spend the night. A troop will not stay in the same area for more than a day and will not use the same nest twice. Each morning, the silverback leads its troop to a new area where there is plenty of food. If the group is attacked by humans, leopards, or other gorillas, the silverback protects its troop members even at the cost of its own life.

Gorillas sleep about thirteen hours at night and nap for several hours at midday. When they are not resting, they are most likely eating. An adult male will eat up to forty pounds of food a day. Gorillas' diets consist mainly of plant foods such as leaves, shoots, fruits, bulbs, bark, and vines, although they also like to eat ants, termites, and worms.

Gorillas are very intelligent and can communicate with each other by using gestures, body postures, facial expressions, and vocal sounds. They are also able to comprehend spoken languages and can learn to communicate in sign language. Gorillas display a full range of emotions, including joy, grief, love, hate, pride, and shame. They laugh when they are tickled, and cry when they are sad or hurt.

Despite their massive size and strength, gorillas are actually shy and peaceful creatures.

Wacky Fact:
No two gorilla noses are alike! Researchers in the wild take close-up photos of each gorilla's face to help identify individuals.

Water Resistant: Gibbons tend to avoid water, since they do not know how to swim.

Fast Fact: A gibbon's territory includes separate sleeping trees for all family members.

The Gibbon

Gibbons are the acrobatic stars of the animal kingdom, as they are the fastest and most agile of all tree-dwelling, non-flying mammals. They live mainly high up in the treetops and are rarely on the ground. They do most of their traveling through the trees by swinging from branch to branch. This is called brachiating. Gibbons have strong, hook-shaped hands for grasping tree limbs, extremely long arms for reaching faraway branches, and special shoulder joints to give them a greater range of motion when swinging. While they are brachiating, they swing as fast as 35 mph and are able to swing from branch to branch for distances of up to fifty feet. Gibbons also enjoy jumping from treetop to treetop and are able to leap up to thirty feet in a single jump!

Gibbons have very good bipedal locomotion. This means that when they walk, they often do so on two feet, throwing their arms above their head for balance. Their agility allows them to walk on small branches high in the air.

In addition to their acrobatic abilities, gibbons are famous for their ability to "sing." They do this to announce their presence and to warn others to keep away from their territory. Gibbons' songs are usually sung as a "duet" by the adult male and female, with their offspring sometimes joining in. Their calls are quite loud and can be heard up to one mile away.

Gibbons are much smaller and more slender than the other apes, and are classified as "lesser apes," as opposed to one of the "great apes" (chimps, gorillas, orangutans, and bonobos). They weigh only up to twenty-five pounds and are seventeen to thirty-five inches long. Gibbons do not build nests like the great apes. They sleep sitting up, with their arms wrapped around their knees and their heads tucked into their laps. Another difference between gibbons and the "great apes" is that in the gibbon family, the females usually weigh more than the males.

There are sixteen different species of gibbons, ranging from India to China to Borneo. Gibbons come in various colors, including black, gray, brown, and yellow. Many have white markings on their faces, hands, and feet. Their diet consists of fruit, leaves, insects, and small birds.

Did You Know?
Gibbons rarely use physical contact to defend their territory. Instead, they use calls to warn others to stay away.

Shoo Fly: When orangutans are bothered by mosquitoes, they use branches as fly swatters to chase the mosquitoes away.

Fast Fact: In the Malay language, orangutan means "person of the forest."

The Orangutan

Orangutans, considered by many zoologists to be the most intelligent of all animals, have been extensively studied for their learning abilities. Orangutans are able to manufacture an assortment of elaborate tools and use each one for a specific job. Their "tool box" consists of fishing tools to catch fish, insect-extraction tools for use in tree cavities, and seed-extraction tools for taking seeds out of hard-shelled fruit. In less than six minutes, orangutans are able to construct sophisticated sleeping nests from branches and foliage.

They use leafy branches as umbrellas to protect themselves from rain and sun, and sometimes drape large leaves over themselves to use as a poncho!

Orangutans have a unique approach to problem solving. They are slow and deliberate, taking time to carefully work out solutions to problems before putting a plan into action. One example: When a chimp is given an oddly shaped peg and several different holes to try to put it into, the chimp will immediately try shoving the peg into various holes until it finds the right hole for the peg. But an orangutan will handle this test quite differently. It will stare off into space, and then, after a short while, will carefully place the peg into the correct hole. Orangutans have also been known to watch villagers use boats to cross local waterways, and then untie a boat and ride it across the river on their own!

The orangutan's reddish hair clearly distinguishes it from other great apes. An adult male is four to five feet tall and can weigh up to 220 pounds, while the females weigh about half as much as the males.

Orangutans are the most arboreal of the great apes; they spend most of their time in trees. They travel by swinging from branch to branch with their very long and powerful arms and hook-shaped hands and feet. Orangutans have an enormous arm span; males can stretch their arms seven feet across, from fingertip to fingertip.

Orangutans get much of their food from the trees where they live. More than half of their diet consists of fruit. They also eat nuts, bark, plants, ants, termites, and bird eggs.

Did You Know?
Orangutans love to eat soap, though they'll lather their arms first, before eating it!

Chimpanzee

In the wild, female chimpanzees give birth every five or six years. They usually give birth to a single infant. For the first few months of its life, a baby chimp will cling to its mother's belly, even as they travel. Afterward, the baby chimpanzee will ride on its mother's back until it reaches two years of age. Young chimpanzees spend their first seven to ten years learning from their mother how to find food, build nests in trees, use tools, and how to groom. Chimps can live up to fifty years in the wild.

Gorilla

After an almost nine-month pregnancy, a female gorilla gives birth to a single baby. The baby is small, weighing about four and a half pounds. Baby gorillas ride on their mothers' backs from the age of four months through the first two or three years of their lives. At five to six months of age, they learn to walk, and by eighteen months of age, they can follow their mom on foot for short distances.

Young gorillas stay home with their mother until they are four to six years old. They learn the necessary skills of survival by watching their mom and fellow troop members. Gorillas can live up to thirty-five years in the wild.

THE CIRCLE OF LIFE

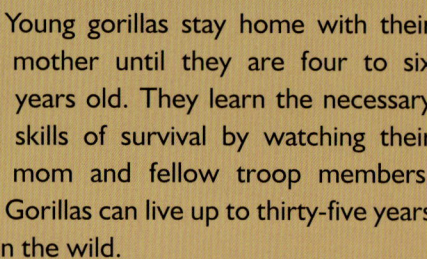

Gibbon

Female gibbons usually give birth every two to three years. In most cases, one baby is born at a time. Newborn gibbons are born with a small cap of fur on the top of their heads. During the first year of its life, the baby gibbon is guarded closely by its mother.

Gibbons become physically independent when they reach three years of age, but they will remain with their parents until they are about five to six years old. At that time, they leave to start a family group of their own. Gibbons can live up to forty years in the wild.

Orangutan

Female orangutans give birth only once every eight years — the least often of any animal. Orangutans will remain with their mothers for about seven years, until they develop the skills necessary to survive in the wild. They have the longest childhood of all the great apes, and for good reason; when they grow up, they don't have a troop around to guide them as do the others. Since orangutans are solitary animals, they need to learn all of life's lessons before they set off on their own. Orangutans can live up to forty years in the wild.

Baby Fact:
Baby orangutans are born in nests.

Torah Talk

"Adnei ha'sadeh is considered an animal. Rabbi Yosi says: It causes spiritual impurity [when dead] in a building, like a human being" (*Kilayim* 8:5).

The above mishnah states that when the creature known as *"adnei ha'sadeh"* dies, it imparts *tumah* (spiritual impurity) on the same level of severity as a dead human (i.e. it can impart *tumah* to anything under the same roof as its carcass). What is this creature that is considered a beast, yet is close enough to a human being that it shares the same laws of ritual impurity?

The Gemara (*Yerushalmi, Kilayim* 8:4) elaborates: "*Adnei ha'sadeh,* the *yaysi araki,* is the mountain-man."

What is the mysterious mountain-man? The answer isn't entirely clear, but Rabbi Yisrael Lifshitz, a nineteenth-century commentator on the mishnah, provides the following explanation in his sefer, *Tiferes Yisrael*:

"It seems to me that it means the 'wildman,' which is called 'orangutan.' This is a type of large ape, genuinely similar to a person in form and build, except that its arms are long, reaching to its knees. It can be taught to chop wood, to draw water, and also to wear clothes, just like a human being; and also to sit at a table and eat with a knife, fork, and spoon. In our times, it is only found in the great jungles of Africa; however, it appears that it may formerly have been found also in the vicinity of the Land of Israel, in the mountains of Lebanon, where even in our day there are great forests, of the 'cedars of Lebanon' fame; and therefore it was called 'the mountain-man.'"

The term *"adnei ha'sadeh,"* explained to mean "men of the field" (see *Encyclopedia Talmudis*), may well also include other great apes. The Malbim (*Vayikra* 11:27) states that it refers to chimpanzees as well as orangutans. The Rambam notes that "those who bring news from the world state that it speaks many things which cannot be understood, and its speech is similar to that of a human being." This description would be adequately accurate for all of the great apes. Of course, it may refer to a different creature that no longer exists. For the time being, the best we have to go on is Rabbi Lifshitz's definition of the orangutan.

Record Holder:
On January 31, 1961, an ape named Ham became the first chimp in outer space. His name is an acronym for the lab that prepared him for his historic mission — the **H**olloman **A**erospace **M**edical Center.

INTERESTING FACTS & STATS

🐾 Chimpanzees can recognize themselves in the mirror. Most animals don't have this ability to reason and will attack a mirror image of themselves.

🐾 Grooming is an important behavior for chimps, both socially and for skin care.

🐾 Chimps sometimes chew leaves to make them absorbent and then use them as a sponge, dipping them in water and sucking out the moisture.

🐾 Gorillas will make their nest either on the ground or in a tree.

🐾 Gorillas can be taught to drink from a cup, eat with utensils, write on paper, and type sentences by pressing symbols on a computer keyboard!

🐾 Gorillas do not beat their chests with their fists, but with open, cupped hands, making their familiar loud sounds which indicate aggression or excitement.

🐾 Gibbons' "songs" vary from species to species, and even males and females of the same species sing different "songs."

🐾 The gibbon has a unique wrist. It is composed of a ball and socket joint, allowing for biaxial movement (movement in two directions).

🐾 The *siamang* is the largest species of the gibbons. It is most famous for its throat pouch, called a *gular sac*. This pouch can be inflated to the size of the siamang's head, allowing the animal to make loud, resonating songs.

🐾 As adult male orangutans age, they develop cheek pads on their faces and pouches on their throats.

🐾 When a baby orangutan following its mother reaches a gap between trees that is too wide for it to cross, the mother will use its body as a bridge for the baby to scamper across.

Ape Trivia

1. **Which is the only great ape not found in Africa?**
a. chimpanzee b. gorilla c. bonobo d. orangutan

2. **What does the word 'gorilla' mean?**
a. the hairy one b. the gory one c. a mighty warrior d. the dumb one

3. **Which ape travels mostly on the ground?**
a. chimpanzee b. spider monkey c. gibbon d. orangutan

4. **Which ape has the longest arms, relative to its body size?**
a. chimpanzee b. gorilla c. gibbon d. orangutan

5. **Which male ape prefers to be by itself, and is considered a loner?**
a. chimpanzee b. gorilla c. gibbon d. orangutan

ANSWERS: 1. d 2. a 3. a 4. c 5. d

Animal Crackers:
Q. What did the banana say to the gorilla?
A. Nothing; bananas can't talk!

BEAR דוב

Believe It or Not: Black bears DO NOT growl. Most sounds you hear them make on video are voice-overs and not from actual bears!

Fast Fact:
Black bears are fast runners and are able to run as fast as 35 mph uphill.

The American Black Bear

The American black bear is the most common bear native to North America. These bears typically live in forests, but are also found in mountains and swamps. The coat of an American black bear is shaggy and usually black, but despite its name, it can also be blue-gray, brown, cinnamon, or even (very rarely) white.

The American black bear is about four to seven feet long, two to three feet high standing on its four legs, and weighs between 150 and 300 pounds. It has small eyes, a long brown snout, rounded ears, a large body, and a short tail.

It is an excellent tree climber, and despite its size, regularly ascends to the treetops to eat. Besides using its sharp claws to climb, it also uses them to rip open old logs in search of grubs and worms. American black bears are omnivores; they feed on many different types of food, including both plants and animals. They mainly feed on vegetation, but they use their great sense of smell to search out fruit, nuts, berries, honey, and roots. They will also eat fish, small mammals, and insects.

American black bears are not true hibernators, but they are considered 'highly efficient' hibernators. They sleep for months without eating, drinking, or having to relieve themselves, but may wake up if disturbed. They enter their dens in October and November. Prior to that time, they can gain up to thirty pounds of body fat to help them get through the seven months during which they fast.

Hibernation in black bears typically lasts three to five months. During this time, their heart rate drops from forty to fifty beats per minute to eight beats per minute. Their dens vary and can be in the form of caves, burrows, under logs or rocks, or even high above the ground inside of trees!

In comparison to true hibernators, their body temperature does not drop significantly (around ten degrees lower), and they remain somewhat alert and active. If the winter is mild enough, they may wake up and forage for food. Females also give birth in February and nurture their cubs until the snow melts.

American black bears are found in the forested areas of the United States, Canada, and Mexico.

Did You Know?

A group of American black bears can be called either a *sleuth* or a *sloth*. However, a group of polar bears is called a *celebration* (no kidding).

Shhh...
A mother polar bear can give birth and nurse her young while still in her winter sleep!

Baby Fact:
Polar bear cubs learn to freeze and remain still while their mother hunts. If they move, the mother disciplines them with a whack to the head!

The Polar Bear

The polar bear roams along shores and on sea ice in the icy cold Arctic. It is the largest bear, and also the world's largest land carnivore. A boar (adult male) can be ten feet tall and weigh between 800 and 1,500 pounds while a sow (adult female) weighs about half of that. Although most polar bears are born on land, they spend the majority of their time at sea. They are very strong swimmers, and their large front paws, which they use to paddle with, are slightly webbed. Polar bears have been known to swim one hundred miles at a stretch. Their eyesight is pretty impressive even underwater; they are able to see their prey from fifteen feet away.

The scientific name for the polar bear is *Ursus Maritimus*, which means "sea bear."

Polar bears are made for the cold, and for good reason; they live in one of the coldest environments in the world. Their hairs are hollow, making for excellent insulation that is capable of trapping much of their body heat, plus their fur and thick skin absorb sunlight for warmth. Additionally, they have a layer of fat which acts as a nutritional reserve when food can't be found, and this layer also helps them generate heat to help insulate them from the freezing air and cold water. Fur even grows on the bottom of their paws, which protects against cold surfaces and gives the bears a secure grip on the ice. As a matter of fact, polar bears are built to stay so warm in their cold habitat that sometimes they overheat, and have to cool off in the chilly water!

Polar bears are meat eaters and primarily eat seals. When sea ice forms over the ocean in cold weather, many polar bears head out onto the ice to hunt seals. The polar bear's nose is so powerful, it can smell a seal on the ice twenty miles away, sniff out a seal's den that has been covered with snow, and even find a seal's air hole in the ice up to one mile away!

Polar bears are patient hunters. They often rest, silent and motionless, at a seal's breathing hole in the ice, waiting for a seal in the water to come up. Once the seal pops up, the bear will attack. Polar bears are also known to hunt by swimming beneath the ice. In addition to seals, they also eat walrus, caribou, beached whales, grass, and seaweed.

Wacky Fact:

The actual color of the polar bear's skin is BLACK — not white! Their hair is transparent, made up of a thick hollow core that reflects light. As the light bounces on the air spaces between hairs, you see white, a reflection of all colors combined.

Gone Fishin'
Kodiak brown bears can catch fifteen salmon in an hour!

Fast Fact: Cubs produce a loud humming sound while nursing, which is believed to help stimulate their mother's milk production. This noise is so loud, it can be heard from outside the bears' den!

The Brown Bear

The brown bear is one of the largest bear species and is the most distributed bear in the world. It lives in the forests and mountains of northern North America, Europe, and Asia. There is only one brown bear species, but there are many subspecies of it, such as the Alaskan bear, European bear, Syrian bear, Kodiak bear, and grizzly bear.

The coat of the brown bear ranges in color, from dark brown to reddish brown to cream toned. Although it resembles the black bear, the brown bear is much larger and stronger.

Male brown bears weigh about 700 pounds, and the females average about 350 pounds. Standing upright on its hind legs, an average-sized male brown bear may reach seven feet tall. Despite their enormous size, brown bears are pretty fast, and can run as fast as 40 mph!

Brown bears are omnivores and will eat anything nutritious that they find. Most of their diet consists of plants and vegetation, but they will certainly eat meat if they can find it. They occasionally hunt prey such as rodents, young deer, and elk. However, the food that first comes to mind when one thinks of bears is salmon. Brown bears excel at fishing and love to eat fish, especially salmon.

Brown bears have a large hump of muscle on top of their shoulders and long, strong, front claws that enable them to dig up roots to eat, tear apart logs in search of grubs, and hollow out dens for hibernation. In autumn, they may eat as much as ninety pounds of food a day, as they prepare for their hibernation, since they will need to live off their stored body fat for up to seven months. They can weigh twice as much before hibernation as they do in the spring.

Brown bears dig large dens for their winter homes and spend four to seven months a year sleeping in them. This sleep is commonly called hibernation, but unlike true hibernation, the bears' body temperature does not drop drastically. However, the bears' metabolism slows down, as does their heart rate, which decelerates from up to seventy beats per minute to only ten beats per minute. The bears' "winter sleep" is what allows them to stay alive for a long period of time when there is little or no food available to them. Bears in warmer climates spend less time in their dens than those in areas that have a longer winter.

Did You Know?
Despite its "large" reputation, the grizzly bear is actually one of the smallest of the brown bear subspecies.

Lite Weight:
A baby panda weighs 1/900 of its mother's weight — making it the smallest mammalian newborn relative to its mother's size!

Fast Fact:
The Chinese name for panda is Da Xiong Mao, which means "large bearcat."

The Giant Panda

High in the cold and rainy bamboo forests, in the misty mountains of southwestern China, lives one of the world's rarest mammals: the giant panda. Unfortunately, the giant panda is an endangered species, as there are fewer than 1,000 of these beautiful creatures remaining in the wild.

A giant panda is often called a panda bear. But not all zoologists agree that it belongs in the bear family. At one time, the giant panda and the unrelated red panda were placed in the raccoon family. Pandas and raccoons have similar markings and possess similar traits, such as using their front paws to hold their food. Now, however, zoologists are able to learn about an animal by studying its genes. Many zoologists agree that giant pandas have similar genes to bears and belong in the bear family.

The giant panda has a body shape typical of bears. It also has a thick, white, wooly coat with black fur on its ears, eyes (forming "eye patches"), shoulders, arms, and legs. This thick fur keeps the giant panda warm in the cold forests.

Compared to most bears, giant pandas are not considered 'giant' and are actually small. But they eat a lot for their size. Giant pandas spend at least twelve hours each day eating! They rarely eat anything but plants, and bamboo is the most important plant on their menu. Because bamboo is so low in nutrients, pandas eat as much as eighty-five pounds of it each day. Occasionally, they will eat other vegetation, fish, or small animals, but bamboo accounts for 99 percent of their diet.

Pandas' molars are broad and flat. The shape of their teeth helps the pandas crush the tough bamboo shoots, stalks, and stems that they need to consume. Pandas are able to grab the bamboo stalks with their front paws, which have enlarged wrist bones that act as thumbs for grasping.

Although the average panda is about five feet in length and weighs between 200–300 pounds, it is quite shy. Pandas do not like to wander into areas where people live, and therefore, if there is no longer bamboo available in their area, they will starve rather than move to new areas where bamboo does exist.

The giant panda is able to climb and take shelter in hollow trees or rock crevices, but does not establish permanent dens. For this reason, pandas do not hibernate and will instead move to elevations with warmer temperatures.

Wacky Fact:
People are unable to tell if a baby panda is a girl or a boy during its first four years!

American Black Bear

Female black bears usually give birth to two cubs every other winter, although their litter can be as large as five. At birth, cubs weigh between 10-16 ounces and are born blind and helpless. They are not able to open their eyes until 28–40 days after birth and they do not begin to walk until five weeks of age. The mother nurses her cubs until spring, and then all emerge from their den in search of food. Cubs will remain with their mother for about two years to learn how to hunt and survive. American black bears live up to thirty-two years in the wild.

Giant Panda

Newborn pandas weigh only three to five ounces at birth. They are rather helpless and need great care by their mother in order to survive. During the first few days after birth, the mother is with her cub constantly and does not leave her den even to eat or drink! When the cub is seven to eight weeks old, its eyes open, and by ten weeks, it starts to crawl. By twenty-one weeks, the cub is able to walk and play. The cub starts to eat bamboo at seven to nine months of age; however, it continues to nurse until it is eighteen months old. At this time, the cub is ready to leave its mother's den and live on its own. Giant pandas live up to twenty years in the wild.

THE CIRCLE OF LIFE

Brown Bear

Brown bears dig dens for winter hibernation, often holing up in a suitable hillside. Female bears give birth during this winter rest, usually to a pair of cubs. The cubs are born almost hairless, toothless, and blind. Although the mother will sleep through the winter, the cubs spend their time nursing, playing, and keeping warm in her fur. The cubs' eyes open when they are six weeks old, and by spring they have grown teeth and thick fur and are able to follow their mother outside the den. Cubs live with their mothers up to three years before living on their own. Brown bears live up to twenty-five years in the wild.

Polar Bear

Pregnant polar bears make their dens in snow banks in the fall, in preparation for their winter stay. They give birth to 1–3 cubs (usually two) in early winter. A newborn cub is blind, toothless, and hairless and is about the size of a rat. The polar bear mother's milk contains thirty-five percent fat, which helps the cubs grow quickly. In the spring, the cubs start exploring with their mother outside their den and learn how to hunt. At about two years of age, they are ready to be on their own. Polar bears can live up to twenty-five years in the wild.

Baby Fact:

Brown bears can climb trees to eat or escape predators, but only when they are cubs! As they become adults, they become too heavy for climbing.

Torah Talk

In the Gemara (*Bechoros* 8b), a story is related about how Rebbi Yehoshua ben Chananyah once entered into an argument with the Roman Caesar. Rebbi Yehoshua claimed that the Torah scholars were wiser than the philosophers of Athens, while the Caesar believed the opposite. To prove his point, Rebbi Yehoshua wagered that he could trick the philosophers of Athens into leaving their headquarters in Athens and come to the Caesar's palace in Rome. Rebbi Yehoshua traveled to Athens and, after an intriguing debate with the philosophers, he succeeded in his mission.

When he presented the philosophers before the Caesar, however, the Caesar did not believe that the men Rebbi Yehoshua brought before him were the genuine philosophers. These men lacked the arrogance for which the philosophers were famed, and showed instead a meek demeanor. Rebbi Yehoshua had anticipated this, knowing that the philosophers would feel humbled as long as they were away from their home. He took a handful of earth that he had brought with him from Athens and threw it in the air. The philosophers, sniffing the familiar smell of their local habitat, were immediately able to regain their composure and their customary arrogance. Apparently, a person feels timid when not in his natural environment.

Another source for this observation may be found in *Sotah* (47a), where the Gemara discusses the episode that occurred in *Melachim II* (2:23, 24). Hashem produced two bears that came out of a forest and attacked a group of malicious youngsters who had taunted the prophet Elisha. The Gemara explains that it was actually a double miracle that was performed, for not only were there no bears beforehand, but there was also no forest in that place before Hashem brought it forth to avenge the shaming of Elisha.

Why, asks the Gemara, was a double miracle necessary? Why did Hashem create a forest where there was none? Was it not sufficient for Him to create only the bears, without the forest? The Gemara answers that the bears without the forest would not have sufficed. Bears are not violent when they are not near their forest, and the bears would not have attacked the youths had the forest not been there. Even a bear, when it is not in its familiar surroundings, becomes timid.

We learn from here that both humans and animals lose their natural aggression when away from their home environment.

Record Holder:
The largest bear ever recorded was a Kodiak bear, a subspecies of the brown bear, which weighed over 2,500 pounds and was almost fourteen feet tall!

INTERESTING FACTS & STATS

- The most highly developed sense of the black bear is its sense of smell. It is able to sniff out a dead animal twenty miles away!
- The black bears' sense of hearing is twice as strong as humans.
- The most famous black bear lived in Louisiana. It was a black bear cub that game-hunter President Theodore ("Teddy") Roosevelt refused to shoot. This was the inspiration behind the stuffed toy known today as the "teddy bear."
- The only parts of the polar bear that radiate heat are the eyes and nose. To retain that heat, polar bears cover their eyes and nose with their fur-covered paws when they sleep.
- The United States, Canada, Denmark, Norway, and the Soviet Union signed an agreement in 1973 to protect polar bears.
- Polar bears are the fastest four-legged swimmers.
- Brown bears are often called grizzly bears, because the tips of the hair on many of them are grayish, or grizzled.
- Due to the fact that brown bear cubs stay with their mothers for some two and a half years after they're born, the female brown bears only reproduce once every three years.
- Grizzly bears can run as fast as the average horse!
- Giant pandas are very flexible and like to do somersaults.
- The giant panda's tail measures at four to six inches, making it the second longest tail in the bear family (the Sloth bear has the largest).

Bear Trivia

1. Which bear is the largest predator on land?
   a. black bear **b.** polar bear **c.** brown bear **d.** giant panda bear

2. Which bear always has black skin?
   a. black bear **b.** polar bear **c.** brown bear **d.** Kodiak bear

3. Which bear is the rarest to find?
   a. black bear **b.** polar bear **c.** brown bear **d.** giant panda bear

4. Which adult bear can swim long distances AND is able to climb high atop trees?
   a. black bear **b.** polar bear **c.** brown bear **d.** giant panda bear

5. Smokey Bear is the mascot of the U.S. Forest Service. What type of bear is he?
   a. black bear **b.** polar bear **c.** brown bear **d.** grizzly bear

ANSWERS: 1. b 2. b 3. d 4. a 5. a

Animal Crackers:
Q. What's the difference between a polar bear and a panda?
A. About 1,500 miles

The Beaver

Beavers are the largest rodents in North America and the second largest in the world, after the South American capybara. Beavers are able to live on land and in water.

Beavers move with an awkward waddle on land, but are graceful in the water, where they use their large, webbed rear feet like swimming fins, and their paddle-shaped tails like rudders. These attributes allow beavers to swim at speeds of up to five miles an hour. As the beaver dips underwater, its nose and ears shut to keep water out. Beavers can remain underwater for fifteen minutes without surfacing, and have a set of transparent eyelids that function much like goggles. Their fur is naturally oily and waterproof.

Beavers are famously busy, and they turn their talents to re-engineering the landscape as few other animals can. Knocking down trees with their strong teeth and powerful jaws, they create massive log, branch, and mud structures to block streams and turn fields and forests into the large ponds that they love. This is often very beneficial to humans, as dams built by a colony of beavers help slow the flow of floodwaters and help control erosion.

Adult beavers usually weigh 35–60 pounds and are 3–4 feet long. Males and females look alike, both having dark brown fur. Beavers have very bad eyesight, but they make up for this with their excellent senses of smell, touch, and hearing.

There are two species of beavers: American and Eurasian.

The American beaver can be found throughout North America, except for the most northern parts of Alaska. This beaver's dense, yellowish-black, black, or reddish-brown fur retains body heat even in the coldest water. A secretive and nocturnal creature, it is awkward on land and thus vulnerable to predators.

Eurasian beavers were nearly hunted to extinction in Europe, both for fur and for *castoreum*, a secretion of its scent gland believed to have medicinal properties. However, the beaver is now being re-introduced throughout Europe. Unlike the American beaver, the Eurasian beavers don't always build lodges. Sometimes they dig burrows in stream banks; however, the entrance is also underwater.

Did You Know?
Beavers have bright orange teeth!

THE CIRCLE OF LIFE

The Circle of Life

Beaver families can have as many as ten members in their lodge, and they typically consist of two litters and the two parents. A family of six or more beavers all living together in a lodge is called a colony.

Beaver babies are called *kits*. They are born in litters of about three to four kits. They are born with a full coat of fur and with their eyes open. Within twenty-four hours, they can swim, and after several days, they are able to dive and explore some of their surroundings with their parents. The kits spend their first month in their safe, warm lodge. Their mother is the primary caretaker, while their father maintains the territory. During this time, they learn many important skills through imitation and experience. Yearling beavers (beavers that are between one and two years old) also help care for new kits born in the lodge.

Older offspring, which are around two years old, may also live in families and help their parents. In addition to assisting to build food reserves and repairing their dam, two-year-old beavers will also aid in feeding and protecting younger offspring. While these two-year-old beavers help increase the chance of survival for younger offspring, they are not essential to the family, and only stay to help out in times of food shortage, high population density, or drought.

Adult beavers leave their parents' lodge when they are about three years old. They set up their own territory and find a mate. When beavers leave their territories, they usually do not settle far. Beavers can recognize their kin by using their keen sense of smell. Being able to recognize kin is important for beaver social behavior, and it causes more tolerant behavior among neighboring beavers.

Beavers usually interact and share territory only with their family group. Most beavers live ten to twenty years in the wild.

 What's for Supper?

Beavers are herbivores (vegetarians). They don't eat any meat. They prefer to eat bark, twigs, roots, and herbaceous plants, such as clover, raspberry canes, and aquatic plants. The beaver's chisel-like front teeth grow continuously to counter intense wear brought on by chewing their favorite foods. Trees provide the beaver's favorite winter food — bark and leaves. In summer other vegetation, especially aquatic plants, berries, and fruit make up their diet. The American beaver's favorite food is the water lily.

Baby Fact:
A newborn beaver in its first year of life is called a kit, while a young beaver in its second year is called a yearling.

Unique Traits

The beaver has so many unique features, but none are as extraordinary as its teeth. A beaver has twenty teeth, including the four strong, curved front teeth called incisors. These long front teeth are made with almost unbreakable enamel and are used for gnawing through the trunks of trees.

The beaver's logging operations are as amazing as its carpentry and engineering. Standing on its hind feet, it eats around a tree until the tree looks like an apple core — with a base on bottom and an upper part, and only a very thin, almost toothpick-sized piece of trunk connecting the two segments. The shaky tree is bound to fall; it's just a matter of time before either the wind or the law of gravity finally brings it crashing down.

It is the icy northern winter that makes a beaver build a dam. Winter means no open water to plunge into for refuge, and snow is difficult to travel through to find bark to eat. So our industrious beaver creates a personal pond, in whose muddy base it can anchor a whole winter's supply of timber to eat, and on which it can build an impregnable mansion for itself and its family. The beaver begins by felling a tree near a river, causing it to jam near the point where it intends to build. Once set, the tree catches driftwood, and the beaver furiously lugs in more material from the riverbanks — mud, sticks, stones, and grass, working it all into one big mass, to consolidate the structure. Mud is the major ingredient, and the beaver works it into place with its hands and the side of its face.

How wide is the dam? Think big — it may be as long as 2,000 feet! Once the dam is completed, and the water level established, the beaver then builds a home for its family. The lodge's foundations are sticks, stones, and twigs, woven so professionally that it cannot dissolve or collapse. There are two entrances (both underwater); one acts as an escape hatch if submersible enemies enter to pay an unexpected call! The top of the lodge, made of heavily woven thatch, is not completed until freezing weather sets in. Then, the beaver plasters the roof thickly with mud, which freezes into an armor plate often ten inches thick. None of the beaver's predators have the strength to tear through that roof.

The fact that beavers are sociable, industrious, and faithful to their families, and that they are fully equipped to chop down trees and build solid dams and intricate homes, is an inspiration. Mr. Beaver can thank the Creator Who designed and formed him with such wisdom. And we, humans who observe it all, can use the minds with which we were created, to gain inspiration and strengthen our *emunah*.

Wacky Fact:
Beavers can close both their ears and nose!

Torah Talk

The Gemara in *Gittin* (57a) tells of the conversion of Onkelos, the nephew of a Roman emperor. Undecided as to whether or not to proceed with the process of becoming a Jew or remaining a gentile, he summoned the soul of Bilam and asked him, "Who are the ones that are important and prestigious in the next world?"

"Yisrael," was the quick reply.

"Should I join them?"

"Don't seek their peace or their good ever!" replied the spirit of Bilam.

It is quite shocking. A man living in the Next World, the World of Truth, sees the greatness of our people, and yet advises Onkelos not to associate with them. How could Bilam give such poor advice? Rav Chaim Zichik *zt"l* proposes that the answer lies in the fact that when one passes on to *Olam Haba*, one takes with himself the values and traits he had developed here in this world. Bilam, who groomed a real hate for the Jews in this world, could not abandon those negative feelings even when residing in the World of Truth.

In his sefer *Kol Dodi*, Rav Shalom Schwadron *zt"l* points out that many people work on repentance throughout the month of Elul and through the Aseres Yemei Teshuvah and yet return to their sinful habits soon after the High Holy Days pass.

To explain this phenomenon, he provides some insight into the nature of a beaver. The beaver's instinct is to walk in a straight line. It is not able to circle past a trap even though its life depends on avoiding the hole in its path. The beaver stands at the edge of the trap, whines a bit — and then moves forward right into the clutches of the hunter's lair.

This may explain the saying of Resh Lakish (*Eruvin* 19a): *"The wicked, even at the doorway of Gehinnom, do not repent!"* Stubborn in their ways, they do not improve or repent even when faced with the punishments that await them.

Breaking a habit is one of the most difficult endeavors a man can attempt. We learn from Bilam (and the beaver) that if one does not accomplish self-improvement here on Earth, then one takes his or her flaws with oneself to Eternity. Therefore, one should work on his flaws and try to improve himself, so that one truly dies the "death of the righteous" and goes to the Next World (after a full, good life in this world), ready to enjoy the World of Truth. To die like a Jew, one must live like a Jew.

Record Holder:
The largest dam ever built by a beaver is 2,790 feet long, which is more than twice the length of the Hoover Dam and can be seen from space!

INTERESTING FACTS & STATS

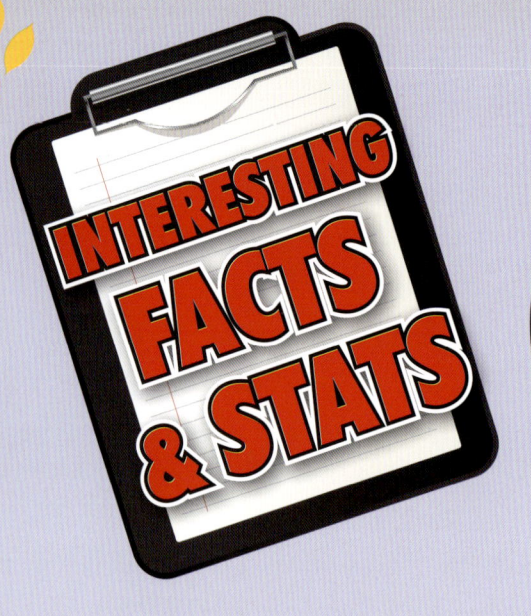

🐾 Beavers do not hibernate, and lodges need to be built in deep enough water so the entrances do not freeze in the winter.

🐾 Beavers mark their territory with "scent mounds" — piles of mud and sticks that the beaver coats in musk oil.

🐾 In autumn, beavers store gnawed wood inside the reservoir near their lodge. This place is called the "storehouse."

🐾 Beavers are second only to humans in their ability to manipulate and change their environment.

🐾 On land, the beaver's tail helps the beaver sit up. In water, it acts like a rudder.

🐾 Transparent inner eyelids close over each eye to help the beaver see underwater.

🐾 The beaver waterproofs its thick fur by coating it with castoreum, an oily secretion from its scent glands.

🐾 About half of all American beavers have pale brown fur; 6 percent have black fur and the rest are other shades of brown.

🐾 Adult beavers have long flat tails that are about a foot long.

🐾 A beaver can remain submerged in the water for up to fifteen minutes.

🐾 There were once more than 60 million beavers in North America. The numbers have been reduced significantly due to humans hunting them for their fur and glands (used in medicines and perfumes).

Beaver Trivia

1. **A beaver in its second year of life is called a…**
a. kit b. pup c. yearling d. bucky

2. **Why do beavers slap their tails in the water?**
a. to catch insects b. to bathe c. to cool off d. to warn each other of danger

3. **Which one of these statements is false?**
a. Beavers have good eyesight. b. Beavers have good hearing. c. Beavers have a good sense of smell.

4. **In the winter, the beaver mainly eats…**
a. carrots b. bark c. flowers d. grasshoppers

5. **What is the beaver's home called?**
a. crib b. den c. lodge d. inn

ANSWERS: 1. c 2. d 3. a 4. b 5. c

Animal Crackers:
Q. What did the beaver say to the tree?
A. It's been nice gnawing you!

CAMEL גמל

The Camel

Camels were domesticated over three thousand years ago, and even today, people depend on them to travel across arid environments. They can easily carry an extra two hundred pounds while walking up to twenty-five miles a day in the harsh desert. A camel is able to travel as fast as a horse, yet is capable of enduring lengthy periods without food or water.

There are two species of camels: *dromedary*, or one-humped camels, which are native to Arabia and other parts of the Middle East; and *Bactrian*, or two-humped camels, which live in Central Asia.

The Dromedary Camel

The dromedary camel, also known as the Arabian camel, exists today only as a domesticated animal. About ninety percent of the world's camels are dromedaries. Even though the dromedary camel has only one hump, it uses it to its optimum effect. The hump stores up to eighty pounds of fat, which breaks down into water and energy for the camel when sustenance is not available. This enables the camel to travel up to a hundred desert miles without water.

The Bactrian Camel

There are two types of Bactrian camels, one wild and one domesticated. Wild Bactrians are very different from domesticated Bactrians. They have smaller humps, less hair, and are thinner.

Bactrian camels have two humps, which function the same way as the hump of their dromedary relatives. But as opposed to having to cope with the shifting sands of the Sahara desert, the Bactrian camels battle the rocky deserts of Central Asia instead. Temperatures in these areas can reach over 100°F in the summer, and drop to −20°F in the winter. The camels' thick, shaggy coats protect them in the winter and fall away as seasons change and temperatures rise.

Similarities

The dromedary and Bactrian camels have many things in common. They were created not only to survive, but rather to thrive, in the harsh elements of the desert. Both types of camels rarely sweat, helping them conserve fluids for long periods of time. They are able to close their nostrils to keep out the desert sand, and they have thick eyebrows and long eyelashes to protect their eyes. Big, tough lips enable them to pick at dry and thorny desert vegetation, while their large, thick footpads help them travel on the rough, rocky terrain and shifting desert sands. As you can see, camels are specially designed to withstand the extremely brutal conditions of the desert.

Did You Know?
The two humps of a Bactrian camel weigh up to 75 pounds. each and form a natural saddle for riding.

THE CIRCLE OF LIFE

The Circle of Life

After a 12–14-month pregnancy, the mother camel gives birth to a single calf. In rare cases, it may give birth to twins. The mother camel will usually give birth to one calf every two years, for a total of about eight camels throughout her life.

A newborn camel weighs eighty pounds at birth and is born without a hump. The calf is born with its eyes open and is able to stand up and run after it is only a few hours old. At birth, the calf is usually pure white in color.

The calf is nursed by its mother until it is big and strong enough to survive on its own. When "calling" its mother, the calf makes "baa-baa" noises like a lamb. The mother camel stays together with her calf for several years.

An adult camel can weigh up to 2,000 pounds. It can grow to be seven feet tall, and the hump itself can give it an additional thirty inches of height. Camels are usually dark brown to dusty gray in color; however, white camels do exist, too.

When camels walk, they move both legs on one side of their body first, and then the legs on the other side, just like giraffes. Their feet are flat and wide; both of these features help prevent them from sinking into the sand. Camels' long necks enable them to reach high branches and also to lower their heads to eat grass and drink water without having to bend their legs.

Camels can live up to fifty years in the wild.

What's for Supper?

Camels prefer to eat grass, grain, salty plants, and dates. Although the camel is technically an herbivore, its diet is not strictly vegetarian. When food is scarce, it becomes an omnivore, making a meal out of anything it can find, including bones, fish, and meat.

Camels are very shrewd when it comes to finding food in their difficult desert environment. Each part of a camel's split upper lip moves independently, which provides its mouth with maximum maneuverability. This enables the camel to get really close to the ground and eat the short grasses that are too difficult for other animals to reach. Camels can go for a week or more without water, and they can last for several months without food. In the winter, camels eat plants that contain enough moisture to sustain them for a few weeks without water.

Baby Fact:
Baby camels are born without a hump, because the layer of fat cannot develop until the camels eat solid food.

UNIQUE TRAITS

Unique Traits

While the beard of a man grows on his upper lip and chin, no hair ever grows out of a person's lips — and for good reason! Imagine the discomfort of bristles at the entrance to your mouth! However, there is one creature that welcomes hair on its lips, and that is the camel. The camel needs this hair there because of its diet. It feeds on twigs, shrubs, dry grass, and thorny scrub, all of which grow in the sand and which can be covered with sharp brambles and thorns. The thick, coarse hairs on the camel's lips provide the perfect protection to prevent the camel from cutting its mouth on its food.

One hump or two?

If you would be asked to draw a caricature of a camel, it would certainly include a hump or two. After all, a hump is to a camel what a trunk is to an elephant. A camel uses its distinctive hump as a pantry. When the camel eats, it transforms the food into fat, which it stores in its hump.

The camel has the ability to store food and water; however, the hump is not, as many people believe, filled with water. Water is stored in cells all over the camel's body, while the hump simply stores fat for the camel. During periods when food is scarce, the camel has the ability to change this fat back into "food" for itself. For this reason, a camel can go up to ten days in the hottest weather without drinking, and even longer without eating. However, after such a time, the camel's "pantry" becomes depleted, and the hump sags and looks empty. When the camel finally does get to drink, it will slurp up more than thirty-four gallons of water at one time! After the camel's enormous drink and feed, the hump soon regains its previous size.

Knowing a camel's drinking capabilities, we should now have even more admiration for our matriarch Rivka, for the phenomenal kindness she displayed in fetching enough water for Eliezer and his ten camels.

The Bactrian Camel

Wacky Fact:
Camels have three eyelids. Two have long eyelashes and the third is transparent and allows the camel to see during a storm — even when the eyelid is closed!

TORAH TALK

Torah Talk

The camel is mentioned numerous times throughout Tanach. Indeed, the camel has been of service to mankind in a variety of ways for a long time. Its hair was used for the making of clothes, as noted in *Menachos* (39b) and in *Shabbos* (27a), where there is an allusion to the "wool of camels" and we are reminded that it was not to be mixed with sheep's wool (*Kilayim* 9:1). Camel hair was also used for the manufacture of tents, saddle bags, and sandals, because of its durability.

The Gemara even used the camel in the following parable to teach us how to act properly when traveling on the road. (Reckless and impatient drivers should heed the following advice offered by our Sages of old!)

Two camels are attempting to ascend Beis Choron (a dangerous road that cut through a rock in a zigzag fashion). They can traverse the road in either one of two ways.

"If they both ascend at the same time, they will tumble down into the valley below. If they ascend after each other, both can go up safely. How should they act? If one is laden and the other unladen, the unladen camel should give way to the former. If one is nearer to its destination than the other, the one that is nearer should give way to the latter. If both are equally near to, or far from, their destination, they should make a compromise between them. The one which is to go forward compensates the other which has to give way" (*Sanhedrin* 32b).

It is also interesting to note that a camel was involved in choosing the burial place for the Rambam. The Rambam lived and died in Egypt, but requested that his last remains be buried in the Holy Land. When the time for burial arrived, his body was placed on a camel and transported to the Land of Israel. When it reached Teveria, the animal stubbornly refused to go further. The authorities therefore had no alternative but to bury the Rambam in a plot of ground there. However, the grave was not in a deserted or isolated spot, but close to the *kever* of Rabbi Yochanan Ben Zakkai.

The Dromedary Camel

No Sweat! Camels rarely sweat, even in desert temperatures that reach 120°F!

Record Holder:
More than 550 camels competed in a five-kilometer, cross-country race on November 20, 2011 in Inner Mongolia, setting a new *Guinness World Record* for the world's largest camel race.

INTERESTING FACTS & STATS

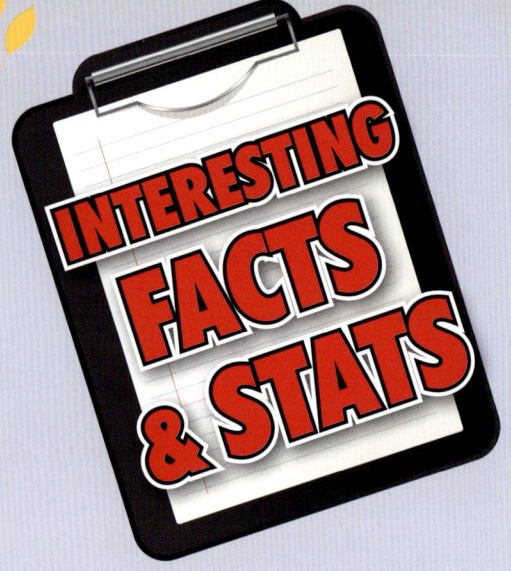

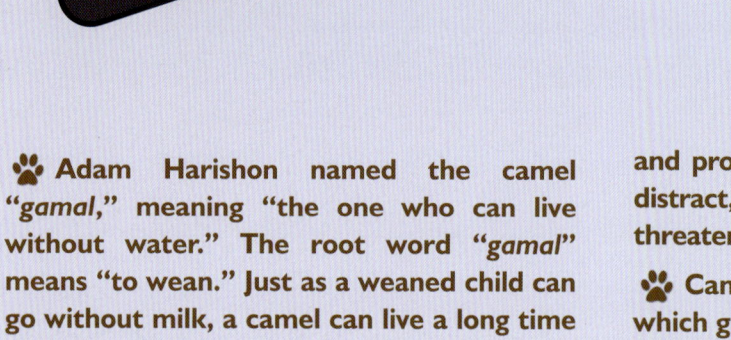

🐾 Adam Harishon named the camel "*gamal*," meaning "the one who can live without water." The root word "*gamal*" means "to wean." Just as a weaned child can go without milk, a camel can live a long time without drinking.

🐾 Camels can kick in all four directions with each of their legs. They will kick when threatened.

🐾 Camels are fast runners and can reach 40 mph in a short burst. They can cruise along at 25 mph when running long distances.

🐾 Camels have a bad reputation of spitting, but they do not actually spit. What they are in fact doing is vomiting! They bring up the contents of their stomachs, along with saliva, and project it out. This is meant to surprise, distract, or deter whatever the camel feels is threatening it.

🐾 Camels have large eyes and big nostrils, which give them good sight and smell.

🐾 Camels can moan, groan, bleat, bellow, and roar. They also make a rumbling growl.

🐾 Camels are often referred to as "ships of the desert," for a couple of reasons:

1. They carry freight and people across long distances like a ship at sea.

2. When camels gallop, both legs of the same side rise and fall together. This action produces a swaying, rocking motion that makes some riders "seasick."

Camel Trivia

1. How many compartments does a camel's stomach have?
a. two **b.** three **c.** four (like a cow's)

2. Which one of these statements is false?
a. Camels are mammals. **b.** Camels store water in their hump. **c.** Camels eat fish.

3. A good indication that a camel has not eaten for a while is when…
a. The camel spits up. **b.** The camel kicks and groans. **c.** The camel's hump droops.

4. What is a male camel called?
a. a colt **b.** a stallion **c.** mister **d.** a bull

5. How many toes does a camel have on each foot?
a. two **b.** three **c.** four **d.** five

ANSWERS: 1. b 2. c 3. c 4. d 5. a

Animal Crackers:
Q. What did the mosquito say when he saw a camel's hump?
A. Wow, did I do that?

COW   פרה

The Cow

The cow is the adult female animal of the cattle family. The bull is the adult male of the cattle family, while a heifer is a female cow that has not yet had a calf. Cows are ruminants, which are animals that chew their cud.

A cow chews its cud for up to eight hours each day. Cows have four digestive compartments in their stomachs (they do not have four stomachs). Ruminants use their multiple stomach compartments to break down food materials with the help of enzymes and bacteria. The partially digested material is then regurgitated and sent back to the mouth, where it is chewed and swallowed again, in order to break down the material even more.

Cows are raised in many different countries around the world, mainly for their natural resources such as milk, meat, and leather. There are three types of cattle in the United States: dairy cattle, which are developed to produce milk; beef cattle, which are raised for their meat; and dual-purpose cattle, which are raised for both milk and meat. Dairy cows provide 90 percent of the world's milk supply. The two main breeds of dairy cows are the Holstein and the Jersey cow.

Of all the dairy breeds in the United States, the Holstein cow is the most numerous; 93 percent of the country's dairy cows are Holsteins! Holsteins are quite large, and sport color patterns of black and white or red and white. A mature Holstein cow weighs about 1,500 pounds and stands fifty-eight inches tall at the shoulder. Its milk has the lowest percentage of butterfat. The Holstein's popularity stems from its ability to produce more milk than any other breed. Top producing Holsteins milked three times a day have been known to produce over 72,000 pounds of milk in 365 days.

A little over 5 percent of the dairy cows in the United States are Jersey cows, making them the second largest dairy breed in the country. The Jersey cow originated on the Island of Jersey, a small British island in the English Channel off the coast of France; hence its name. This cow tends to be a golden-brown color, with a black nose and black hooves. Jersey cows are the smallest of all dairy breeds. A mature Jersey cow has an average weight of 900 to 1,200 pounds. It has the richest milk, with the highest percentage of butterfat and protein.

The average cow in the United States produces fifty-three pounds of milk per day, which is equivalent to about 6.2 gallons.

Did You Know?

A cow must have a calf in order to produce milk.

THE CIRCLE OF LIFE

The Circle of Life

The average cow is two years old when it gives birth to its first calf. The calf weighs between eighty to one hundred pounds at birth, and is usually able to stand on its own when it's just a few minutes old.

A calf is born without a functioning immune system and needs its mother's milk within its first thirty minutes of life in order to survive. This is because the mother cow's milk for her calf, called colostrum, contains not only food, but maternal antibodies that will protect the calf against all the common infections it is likely to encounter in its early life. Within twelve hours, the calf will require approximately six liters of this highly specialized milk from its mother.

During its first few days of life, the calf is kept hidden by its mother, since it is not yet able to keep up with the rest of the herd. The cow will return to its calf a few times during the day in order to nurse it. By the end of the week, the calf will come out of hiding, but it will stay close to its mother. It will continue to remain at its mother's side at all times until it has been weaned (at about six to ten months of age).

A domestic cow can live for up to twenty years.

What's for Supper?

Cows spend their days in herds of around forty to fifty cows, grazing on the grasslands and shrubbery. A cow is able to digest materials that are difficult and nearly impossible for many other animals to digest. Their diet includes grass, hay, unprocessed grains, and chicken litter. Hay and grasses are especially rich in cellulose, a substance that is very hard to digest. Cows, however, have anaerobic bacteria (bacteria that grow in places with little or no oxygen) in their digestive tracts, and these bacteria enable them to digest the cellulose.

It is quite fascinating how the cow is able to use the grass and shrubbery that it eats and convert them into two food products that have some of the highest protein and nutrient value known to man: meat and milk.

Baby Fact:
The process of giving birth to a calf is known as *calving*.

UNIQUE TRAITS

Unique Traits

The cow's udder is a most remarkable "factory," capable of the most breathtaking feats. Each udder is composed of four quarters, with 60 percent of the milk being produced by the hind quarters. The udder contains special cells, called alveolar cells, that are responsible for synthesizing and secreting the various ingredients necessary for milk production. Once the milk has been made, it is then discharged into ducts, sinuses and cisterns, where it remains until the cow has been milked.

What is a cow's milk made from? The simple answer to this is: blood. (Approximately 500 liters of blood must flow through the cow's udder for each liter of milk that the cow produces!) However, a closer look at the milk production process shows the complexity of it, as numerous components in the cow's blood become key players in providing us with the nutritious and delicious drink called milk.

Milk fat is made up of fatty acids. About half of these fatty acids come from the fats in the cow's diet. The other fatty acids are derived from a combining process in the udder of various components in the cow's blood. The lactose in milk (milk sugar) is derived mainly from blood glucose, while the milk protein is derived primarily from the amino acids present in the blood. Other "ingredients" in milk that are obtained from the blood are calcium, phosphorus, magnesium, potassium, sodium, chlorine, vitamins A and B complex, as well as vitamins D, E, and K!

To say that the cow simply (simply!) transforms grass into milk is a massive understatement. The cow is a virtual chemical factory, taking in grass, digesting it, and then converting all the many ingredients and components of its food into blood, which only then undergoes the awesome change of becoming pure white milk. And all the while, the cow just continues to contentedly munch its grass, absolutely unaware of the wonders and complexities taking place within it!

One Way!
Cows can go up stairs, but not down.

Wacky Fact: Cows sweat through their noses!

Torah Talk

Moshe turned and descended the mountain with the two Tablets of the Testimony in his hand… As he approached the camp, he saw the calf and the dancing (Shemos 32:15, 19).

The Golden Calf that the Jews served in *Chet Ha'egel* symbolized denial of the realities of life. A calf represents playfulness and youth; the throwing of oneself into the pleasures of physical life without worrying about the reality of tomorrow. Gold, on the one hand, represents the sparkling beauty of the physical world; the glitter of life in *Olam Hazeh*. On the other hand, however, because as a commodity gold is so stable, and as a material it weathers much, gold also symbolizes eternity. Thus, to put the calf and the gold together, the "Golden Calf" was the representation of eternalizing a childish way of life, a life of indulgence in the here and now, without paying any heed to the future.

The *parah adumah* symbolizes the exact opposite of this, and it was therefore the antidote for the *Chet Ha'egel*. A *parah* is an adult cow, and it is the color red, which represents physical life that is transitory. The *parah adumah* thus represents a sobering look at the vulnerability of mortal man. Because of this, internalizing the message of the *parah adumah* is the first step in helping to remove the *tumah* of *tumas meis*.

The Brisker Rav taught that the *parah adumah* is an important key to the Final Redemption, which will close the chapter on *galus Edom*. The midrash says that when Eisav (Edom) sold the *bechorah* to Yaakov for food, he gave away his portion in the World to Come (*Midrash Rabbah, Parshas Toldos, perek 63*). But that didn't interest Eisav at all, because his approach to life was the "Golden Calf" approach: to be concerned only about the here and now, *Olam Hazeh*. In the final exile in which we find ourselves today, *galus Edom*, the Jew is often found living according to this way of life, too, and we pray for the Final Redemption to come and free us from this sorry state.

Record Holder:

In 1998, a cow named Lucy set the world record for milk production by a single cow. Lucy produced 75,275 pounds of milk in 365 days. This is equal to twenty-four gallons of milk a day for an entire year!

INTERESTING FACTS & STATS

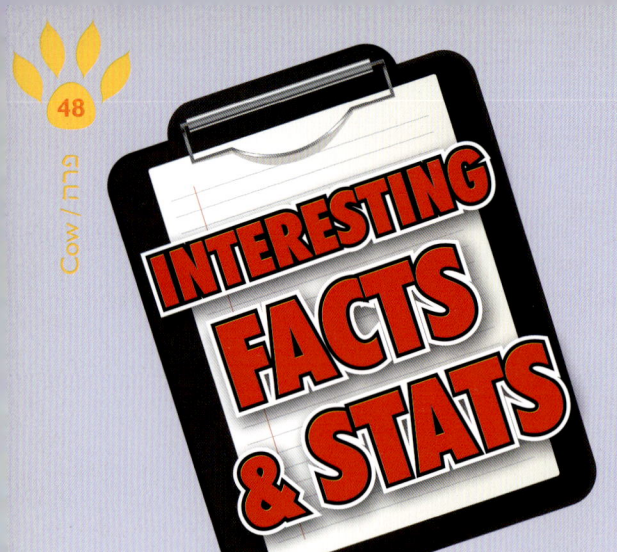

🐾 It used to take a person one hour to milk six cows by hand. Today, with modern machinery, a person can milk one hundred cows in an hour.

🐾 Cows have an amazing sense of smell—they can smell something up to six miles away!

🐾 The yellow color of butter comes mainly from the beta-carotene found in the grass that cows eat.

🐾 Most cows chew at least fifty times per minute.

🐾 An average dairy cow eats about forty pounds of food a day, and drinks about thirty-five gallons of water a day.

🐾 A cow gives nearly 200,000 glasses of milk during its lifetime.

🐾 To make nine gallons of milk, a cow must consume eighteen gallons of water.

🐾 Cows graze by curling their tongues around grass and pulling rather than biting with their teeth.

🐾 A Holstein cow's spots are like fingerprints. No two cows have identical spots.

🐾 It is commonly thought that cows (and bulls) get angry when they see the color red. However, this is not true. Cows are color blind and are unable to identify the difference between colors. The common misunderstanding of this came from the days of bull fighting, when matadors were usually seen waving a red flag. In truth, it was the actual waving of the material that excited the cow, not the color of the material being waved.

Cow Trivia

1. **What is a young female dairy cow called?**
a. a bull b. an ox c. a doe d. a heifer

2. **In what year were cows first brought to America?**
a. 1611 b. 1776 c. 1850 d. 1902

3. **Which statement is false?**
a. Cows have a good sense of smell. b. Cows are color blind.
c. Cows do not have upper teeth. d. Cows have four stomachs.

4. **Besides the Holstein and Jersey cows, there are three other main breeds of dairy cows. Which of the following choices is not one of the main breeds?**
a. Guernsey b. Brown Swiss c. Ayshire d. Danish

5. **What are the most common colors of the Holstein cow?**
a. black and blue b. black and white c. red and white d. brown and white

ANSWERS: 1. d 2. a 3. d 4. d 5. b

Animal Crackers:
Q. What is a cow's favorite *tefillah*?
A. *Moosaf!*

ELEPHANT פיל

The Elephant

Elephants are among the most interesting and recognizable creatures on Earth. You may think that all elephants are large and gray, with big ears and a trunk. However, upon closer inspection, you will come to realize that not all elephants look the same.

There are three different kinds of elephants: the African bush elephant, the African forest elephant, and the Asian elephant (some group the two African species together).

Here are a few ways to tell them apart:

African elephants are the largest land animals on Earth. They are slightly larger than their Asian cousins and can be identified by their larger ears that look somewhat like the continent of Africa. Both males and females have visible tusks, and their skin is very wrinkly. Their backs are sagged (swaybacked), and the two tips at the end of their trunks work like two fingers, helping them pick things up.

Asian elephants have smaller ears, and usually only the males of these elephants have visible tusks. Their skin is not as wrinkly, and they only have one "finger" at the ends of their trunks. Their backs are dome-shaped.

Both African and Asian elephants form deep family bonds and live in tight family groups of related females and young elephants. These groups are called *herds*.

The herd is led by the oldest and often largest female, who is called the *matriarch*. She decides when and where the herd will eat, rest, and travel. Herds consist of 8–100 elephants.

Elephants are extremely intelligent animals and have memories that span many years. It is this memory that serves matriarchs well during dry seasons, when they need to guide their herds, sometimes for tens of miles, to watering holes that they remember from the past.

Elephants make many different sounds. Some sounds cannot be heard by people. Recent discoveries have shown that elephants can communicate over long distances by producing a subsonic rumble that can travel over the ground faster than sound travels higher up through the air. Elephants use these sounds to communicate with each other over long distances.

Elephants use their tusks to dig for roots and water, strip bark from trees, and even fight each other. Elephant tusks are made of ivory and grow throughout the animals' lives. Unfortunately, their tusks have gotten them into a lot of trouble. Because ivory is so valuable, many elephants have been killed by people for their tusks. These dealings have become illegal today, but have not been completely eliminated.

Because of their size, adult African elephants have no enemies other than people.

African Elephant

Asian Elephant

Did You Know?

Elephants have the largest brains in the animal kingdom.

THE CIRCLE OF LIFE

The Circle of Life

Having a baby elephant, called a *calf*, is a serious commitment. Elephants have a longer pregnancy than any other mammal — almost twenty-two months. Female elephants usually give birth to one calf every two to four years. At birth, calves already weigh about 250 pounds and stand about three feet tall!

A calf is usually quite hairy, with a long tail and a very short trunk. It uses its mouth to drink its mother's milk, so it doesn't need a long trunk to feed. Calves are clumsy with their trunks at first, but they learn to use them as they grow older.

As is common with more intelligent species, elephants are born with fewer survival instincts than many other animals. Instead, they rely on their elders to teach them what they need to know.

A new calf is usually the center of attention for herd members. Adults and most of the other young will gather around the newborn, touching and caressing it with their trunks. The baby is born nearly blind and, in the beginning, is almost completely dependent on its trunk to discover the world around it.

Elephants within a herd are usually related, and all members of the tightly knit female group participate in the care and protection of the young. After the initial excitement of giving birth to a new baby, the mother will usually select several full-time babysitters for what is known as *allomothering*. An elephant is considered an allomother when she is not able to have her own calf, usually because she is too old to breed or too young. The more allomothers, the better the calf's chances of survival.

Adult females and young remain together and travel in herds, while the adult males generally travel alone or in groups of their own. Once male elephants become teenagers, they leave the herd.

Elephants can live to be about seventy years old.

What's for Supper?

Elephants are herbivores and spend up to sixteen hours a day eating plants. Their diets are highly variable, both seasonally and across different habitats and regions. Elephants are primarily browsers, feeding on the leaves, bark, and fruits of trees and shrubs, but they may also eat considerable amounts of grasses and herbs. As is true for other non-cud-chewing hoofed mammals, elephants only digest approximately 40 percent of what they eat. They make up for their digestive systems' lack of efficiency in the volume that they eat. An adult elephant consumes 300–600 pounds of food a day!

Baby Fact: Just as a human baby sucks his thumb, an elephant calf often sucks its trunk for comfort.

UNIQUE TRAITS

Unique Traits

When you think elephant, you probably think trunk. Nothing is created without design and purpose — especially the trunk of an elephant. An adult African elephant's trunk is about seven feet long! This most versatile organ of the elephant contains some forty thousand muscles (in contrast, think of your own arm with its two muscles), each of which needs the appropriate messages from the central nervous system in order to function. And function it certainly can! It is partly lip and partly nose, and with its two 'fingers' on the tip, it is used as a worker's arm and hand. It has double hoses, one for sucking in, and the other for spraying out water or dust. When an elephant drinks, it sucks as much as two gallons of water into its trunk at a time. Then it curls its trunk under, sticks the tip of it into its mouth, and blows. Out comes the water, right into the elephant's mouth and down its throat.

The elephant uses its trunk to knock down massive trees, and yet, the trunk is so sensitive that it can pluck a single leaf off a branch. It can be as gentle as the tenderest arm, greeting, scratching, or rubbing; and at the same time, it can be a most efficient weapon, strong enough to kill.

Perhaps it is the elephant's sense of smell which is most amazing. At the tip of its trunk are tiny hairs which detect the precise location of a scent in the wind. And an elephant never forgets the scent of man. It has been said that an elephant, using its trunk, can detect the difference between the scent of a white man and the scent of an African man, at a distance of two miles!

An elephant had once been hunted, but survived. Thereafter, it walked with the greatest care, veering to the left or the right, but always returning to cross its trail. There it would stop, sniff the ground with its trunk, and if there were no scent, it would move on. If it scented a native foot on its trail, it would back into cover and watch until it was sure all was safe. If it scented a white man, it would trample the ground in a rage, and take off in a straight line for miles or more! It knew what kind of weapons white men carried.

The great elephant, with its unique trunk, is as designer-made as the largest man-made machine, only so much more efficient, versatile, and gentle. After learning about this amazing creature, look at the whole world — with its millions of examples of the miracles of creation — and at the vastness of the universe, and come to recognize and appreciate the greatness and wisdom of the One Who made it all.

Wacky Fact:
Elephants are able to swim for long distances; they use their trunks as snorkels when they wade in deep water.

Torah Talk

The first mention of the elephant in Tanach is found in the *Navi* (*Melachim I* 10:22) where it states that Shlomo Hamelech had a navy which every three years brought him gold, silver, *shenhabim* (ivory), apes, and peacocks. The word "*shenhabim*" is translated as "tusk of the elephant."

Rabbi Ben Tzion Shafier utilizes the elephant in the following eye-opening parable:

Being Tied to a Peg in the Ground

In parts of Asia, the elephant remains the beast of choice for lugging heavy loads. As part of its workday, an adult elephant will pull logs weighing thousands of pounds through long stretches of forest undergrowth. Yet at night, that same elephant will be controlled by being tied to a small peg in the ground.

While it would be clear to you and me that a 14,000-pound creature can easily break away from the light ropes holding it, the reality is that it cannot. It cannot escape — not because it isn't motivated, and not because it doesn't want to, but because in the elephant's understanding, it just can't be done.

In this part of the world, shortly after birth, the baby elephant is tied to a peg in the ground. At that stage in its development, it weighs only about 250 pounds and isn't strong enough to break the rope that holds it. From that point and onward, every day of its life, the elephant will be tied to that peg in the ground. Even when the animal has reached maturity and will be called upon to lug felled trees weighing over 4,000 pounds, it will remain tied to a small peg. The understanding is firmly fixed in its mind: it can't escape.

Limiting Beliefs

As Rabbi Shafier explains, many times we are tied to pegs in the ground. There are many situations where we don't reach for greatness, because we are contained — not by ropes, but by limiting beliefs that prevent us from breaking away from the habits and lifestyle choices that stunt our growth.

That process of making the right choices shapes us into different people. It starts the process of change. Once that process is engaged, if we continue on it and consistently choose what is right and proper, step by step we will become different people and view life itself from a different vantage point.

While it is true that serving Hashem requires work, there is a part of each person that naturally yearns to do that which is right. There is a part of us that deeply desires a relationship with Hashem. By attuning ourselves to that part, and by using role models who reached such plateaus, we too can reach the dizzying heights of greatness for which we were created.

Record Holder:

On April 8, 2008, entertainer Fan Yang set a world record by placing a forty-year-old, 8,800-pound Asian elephant named Tai in a large soapy bubble. Tai is recorded in the *Guinness Book of World Records* as 'The Largest Land Mammal in a Bubble.'

INTERESTING FACTS & STATS

🐾 The low, resounding calls elephants make can be heard by other elephants up to five miles away!

🐾 Elephants are able to swim for long distances.

🐾 Elephants spend about sixteen hours a day eating.

🐾 The elephant's eyes are small and its eyesight is poor.

🐾 One elephant molar can weigh about five pounds and is the size of a brick!

🐾 The height of an elephant is 8–13 feet at the shoulder.

🐾 Elephants display signs of grief, joy, anger, and playfulness.

🐾 Males can weigh up to 15,000 pounds, and females can weigh up to 8,000 pounds.

🐾 The heaviest elephant on record was an adult male African elephant that weighed about 24,000 pounds!

🐾 *Pachyderm* means "thick-skinned mammal," and this term often refers to both elephants and hippopotamuses.

🐾 Even though elephants have thick skin, the skin is still very sensitive; in fact, it is so sensitive that the elephant can feel a fly landing on it.

🐾 Elephants often spray themselves with water or roll in the mud or dust for protection from the sun and biting insects.

🐾 Elephants' ears are a little like air conditioners. As elephants flap their wet ears on a hot day, the blood flowing through the many blood vessels there becomes cooled. This in turn cools their large bodies.

Elephant Trivia

1. A group of elephants is called a…
a. flock **b.** quandary **c.** herd **d.** pack

2. Why do elephants flap their ears when wet?
a. to communicate with others **b.** to get rid of insects **c.** to cool off

3. Which one of these statements is false?
a. Elephants are good swimmers. **b.** Elephants have good eyesight. **c.** Elephants can't jump.

4. At what age does the male elephant leave his family?
a. 1–3 years old **b.** 4–6 years old **c.** 7–10 years old **d.** 12–15 years old

5. What is the average litter size of the elephant?
a. 1 baby **b.** 2 babies **c.** 3 babies **d.** 8 babies

ANSWERS: 1. c 2. c 3. b 4. d 5. a

Animal Crackers:
Q. How do you stop a charging elephant?
A. Take away its credit cards.

FOX שועל

The Fox

Foxes are members of the dog family. From afar, a fox might appear as a large animal; however, it is actually quite small. Foxes are usually about the same size as a small, thin dog. There are twelve different species of true foxes. Some of the more familiar species include the arctic fox, fennec fox, swift fox, gray fox, and the red fox. All species share similar characteristics. They possess excellent senses of smell and hearing. Most have large ears; a long, narrow snout; and a long, thick tail. Their coats are usually light brown, reddish, gray, or dark brown (except for the arctic fox, which is white in the winter).

The red fox is the most commonly found species of foxes, and is found almost everywhere throughout the world. Red foxes can be found in all of Canada's provinces and territories, making them one of the country's most widespread mammals. In addition to being in Canada, red foxes can be found in the United States, Europe, Asia, North Africa, and Australia. They live in various habitats including forests, grasslands, mountains, and deserts. They also adapt well to human environments such as farms and suburban areas.

The red fox is the largest of its species and varies in length from 35–42 inches, which includes a 10–16-inch tail. The fox uses its long, bushy tail in many ways, including to maintain a steady balance when running and jumping and as a signal to communicate with other foxes. Additionally, when the fox sleeps in the open, it uses its tail as a blanket to keep itself warm.

Red foxes are nocturnal, but it is not uncommon for them to be seen during the day. They are very good night hunters and usually travel and hunt alone. Besides being able to see in the dark, they are able to use their whiskers to feel their way at night and sense what is around them. Unlike other mammals, the red fox is able to hear low-frequency sounds which help them hunt small animals, even when these small animals are underground!

Did You Know?
Foxes are excellent swimmers, despite a general distaste for getting wet.

THE CIRCLE OF LIFE

The Circle of Life

Female red foxes, called vixens, give birth once a year. The size of their litter can be as little as a single pup and can be as large as thirteen pups. The newborn pups are born blind, and they are not able to open their eyes until they're about two weeks old. At birth, red foxes have blue eyes, and their coats are either brown or gray. After about four to five weeks, their eye color changes to amber, and they grow a new red coat. However, some red foxes have coats that are golden, reddish-brown, silver, or even black.

Both parents help take care of the young pups, although during the first few weeks of life, the pups only stay with their mother inside their den. After one month, the pups are weaned off their mother's milk and start eating pre-chewed food. At four weeks of age, the pups begin to briefly leave their den from time to time.

After about seven months, the young red foxes are ready to hunt on their own and leave their parents in search of their own territory. Males may leave their homes sooner, while the females may stay with their parents a bit longer. Even when the female foxes do leave their parents' den, they usually stay close to their birthplace, while males are known to go as far as 150 miles away.

Foxes can live up to about fourteen years in captivity; however, they only live two to four years in the wild.

What's for Supper?

Red foxes are carnivores. They prey primarily on small animals such as mice, rabbits, birds, and fish. They also like the taste of chicken, and have been called "chicken thieves" by many farmers. The red fox, however, does not only eat meat; it also likes to eat plants, birds' eggs, fruits, berries, and insects. If living among people, they will dine on garbage and pet food, as well.

The red fox likes to hunt and gather food at night, and then sleep during the day. Even when it is not hungry, it will continue to hunt and gather food to store away for future use. It will use this food at times when food is scarce and hard to find.

Baby Fact:
Red foxes that come from the same litter can be different colors!

UNIQUE TRAITS

Unique Traits

The fox has the reputation of being tricky and clever; hence the famous expression, "sly as a fox." What makes the red fox such a crafty and cunning creature?

The red fox uses trickery along with a variety of sly movements in order to hunt its prey, hide, and escape from predators. To begin with, the fox's size alone is deceptive. Its full coat, which is made up of long hairs, makes it look much larger than it really is. The average weight of a red fox is only ten pounds.

The fox uses various methods to catch its prey. When hunting for a mouse, the fox will remain still, allowing the mouse to come near it. When the mouse is close enough, the fox will then pounce on it. Other times, the fox will lure its prey by acting in a bizarre manner. It will start jumping, rolling around, and chasing its tail to attract attention. While doing so, the fox is actually inching closer to its unexpecting prey until it is in range to attack. And then there are times when the fox will cleverly hide and wait for the opportune time to sneak up on its unsuspecting prey.

Although the fox is a predator, it has its own enemies and predators, too. In order to avoid detection from them, the fox uses very clever techniques to cover its tracks. For example, it usually holds its tail high when traveling; however, in freshly fallen snow, it may drag its tail, which suggests that the fox is sweeping away its tracks. Also, when a fox runs fast, its feet create two lines of tracks; however, when it trots, it leaves only a single line of tracks — as its hind feet follow the exact placement of its front feet. Another way the fox tries to avoid detection is by leaving a false trail. It will backtrack or travel in circles in order to confuse its predators.

The phrase "sly as a fox" is certainly an accurate expression because of the shrewdness of this creature!

Wacky Fact:

Although the red fox is a member of the dog family, some of its habits are very cat-like as well, and therefore, it is sometimes referred to as "the cat-like canine."

Torah Talk

The fox is mentioned frequently throughout the Gemara. Below is a beautiful midrash in *Koheles Rabbah* on the verse, *As he came forth out of his mother's womb unclothed, so shall he return* (*Koheles* 5:14).

Geniva, one of the *chachamim*, compared man in this world to a fox that has found a vineyard surrounded on all sides by a high fence, except for one small opening. The fox attempted to enter, but found the hole too narrow to squeeze through. For three days he starved himself, until he was thin enough to enter. Once in the vineyard, he indulged himself to such an extent that he regained his weight, and thus could not leave through the opening. The fox had no choice but to fast another three days in order to leave the vineyard. Then he exclaimed: "Oh, vineyard, how pleasant are you and how desirable is your fruit, but of what benefit are you to me, since I depart from you as thin as when I entered!" Such is the fate of man in this world, says Geniva. As man came forth, so he returns.

The fox also plays a role in the famous mishnah, *Be a tail to lions and not a head to foxes* (*Avos* 4:15). In addition to appearing in *Pirkei Avos*, these words are also found in the Gemara (*Sanhedrin* 37a). But each reference has a different connotation.

In *Pirkei Avos*, these words are associated with leadership. The lion is the king of beasts, the leader and head of the animals. The fox, on the other hand, is noted for its sly and cunning nature; it is also considered to be a foolish animal. One should rather be a tail to a lion, a loyal and trustworthy follower to a truly great leader, than be a corrupt and cunning leader himself.

In the Gemara, these words refer to something of an entirely different nature. In the ancient academy of learning, scholars sat in rows, and when the head of the row was promoted, all the scholars in the row moved forward. The proverb "better a tail to lions than a head to foxes" conveys the meaning that it was better to be placed at the tail of the first row than at the head of the second row.

Tick Tock
The fox's hearing is so sharp that it can hear a clock ticking forty yards away!

Record Holder:
The red fox is one of the fastest mammals in the world. It is capable of running speeds of 48 mph.

INTERESTING FACTS & STATS

🐾 Sometimes a fox may catch a mouse when it isn't even hungry. It will just play with it, and then let it go when it becomes bored.

🐾 Male foxes are called dogs or reynards.

🐾 Baby foxes are called kits, pups, or cubs.

🐾 Fox hunting originated in the United Kingdom in the eighteenth century.

🐾 The world's smallest fox is the fennec fox, weighing 3.5 pounds.

🐾 When building its den, the fox makes sure there is more than one entrance, in case of an emergency.

🐾 A fox can easily jump over a fence that is six and a half feet high.

🐾 The fox's tail makes up one third of its total length.

🐾 Foxes use a variety of different sounds — twenty-eight of them! — to communicate with each other.

🐾 Foxes have retractable claws.

🐾 Foxes can adapt to different environments and are able to live in the deepest wilderness, as well as in your very own backyard.

🐾 Foxes have been used successfully on fruit farms to control pests. They help get rid of rodents without harming the fruits.

🐾 Although foxes are similar in some ways to both cats and dogs, they generally avoid contact with these animals.

Fox Trivia

1. What do you call a group of foxes?
a. a skulk b. a tribe c. a troop d. a frenzy

2. What is the average lifespan of a fox in the wild?
a. one year b. two to four years c. five to ten years d. fourteen years

3. Which family is the red fox a member of?
a. the cat family b. the wolf family c. the dog family d. the rodent family

4. Which statement is false?
a. Foxes have a good sense of smell. b. Foxes have a good sense of hearing.
c. Foxes are good swimmers. d. The color of a red fox is always red.

5. Which of the following markings will help identify a red fox?
a. a white tail-tip b. a black tail-tip c. a black stripe on its back d. small ears

ANSWERS: 1. a 2. b 3. c 4. d 5. a

Animal Crackers:
Q: Why was the poor fox chasing his tail?
A. He was trying to make ends meet.

GAZELLE   צבי

The Gazelle

Gazelles are medium-sized antelopes found in Africa and Asia. They are noted for their graceful stride as well as for their beauty. The gazelle has large black eyes, a variety of stripes and markings that complement their tan-buff coats, and beautiful ringed horns. In most species of the gazelle, both males and females have horns. The horns of the male are usually heavy and ridged, curving up and back, then forward and in. The female's horns, if present, are shorter, straighter, and thinner.

Gazelles live in herds, which can consist of as few as ten or as many as several hundred animals. The coloration of gazelles, as well as the open areas in which they live, make them very conspicuous. Also, their horns offer no protection against large predators, and so they rely upon their great speed and agility to avoid capture. Besides being able to run as fast as fifty miles an hour, gazelles also have great leaping abilities. They use a bounding leap, called *stotting* or *pronking*, to avoid predators.

Stotting is a specific gait that involves taking high, stiff-legged jumps by lifting all four feet off the ground at the same time. Stotting, however, actually slows down the gazelle, making it easier for it to be caught by its predator! Zoologists have deduced that stotting is a method used by the gazelle to shows off its health and fitness to its predator. Since most predators prefer to hunt old or unhealthy animals, stotting tells the predator that this animal is strong and healthy so the predator will be better off catching a weaker animal in the herd. It seems to work, since most cheetahs will break off a hunt when a gazelle stots.

Although there are about nineteen different species of gazelles, the two principal species are Grant's gazelle and Thomson's gazelle. The Grant's gazelle lives in a broader range of territory in Africa, while the Thomson's gazelle has a larger population. Both species share grazing ground and often intermingle. It is easy to tell them apart. The Thomson's gazelle is smaller and has a black band stretching from shoulder to hip which divides its tan and white coat coloring.

Did You Know?
The Thomson's gazelle is named after explorer Joseph Thomson, and is also called "tommie" for short.

THE CIRCLE OF LIFE

The Circle of Life

After a pregnancy of about six months, a female gazelle will leave its herd to give birth alone. The mother gazelle usually gives birth to one or two calves. The newborn calf is helpless at birth, and so the mother gazelle hides her newborn in the long grasses of the plains in order to protect it. With its tawny coloring and its ability to remain perfectly still for long periods of time, the newborn calf is almost invisible, despite lying outside in the wide open plains. In this way, the newborn calf stays hidden from its predators for its first three weeks of life.

The mother gazelle leaves the herd to nurse her calf several times a day. When the calf is three weeks old, it is able to run on its own and soon joins the herd with its mother. The calf continues to nurse from its mother's milk for three to six weeks, and then starts eating solid food on its own.

Female gazelles may sometimes stay with their mother for life. Males, however, leave their mother at around six months of age to join a herd of males.

Gazelles are a predominant food source for all of the major predators in Africa. Cheetahs, lions, leopards, hunting dogs, and hyenas prey on young and adult gazelles alike. The gazelle calves can also fall prey to jackals, baboons, eagles, and pythons.

Gazelles can typically live for about twelve years in the wild.

What's for Supper?

Gazelles usually travel to wide, open areas and plains to eat. They are constantly grazing; their diet consists of grasses, shrubs, fresh green leaves, and herbs. During the bountiful rainy season, thousands of animals gather in large groups to feed. Gazelles like to roam with some of the larger animals, such as the wildebeest and zebra, which trample and graze on the taller grasses, thereby making it easier for the gazelles to feed on the short grass.

Although gazelles prefer foods that are easy to digest, they are ruminants, and they will swallow large, unchewed amounts of hard-to-digest cellulose material during feeding. They will then retreat to a safe area in order to chew their cud and slowly digest this food.

Baby Fact:
A baby gazelle can be called a calf or a fawn.

Unique Traits

Unique Traits

Gazelles live in very arid areas where, at times, water is extremely limited. Their habitats include the dry areas of Africa, Asia, the Middle East, and Siberia. One of the most unique characteristics of the gazelle is its ability to live in hot and mostly dry environments without having the availability of surface water. As a matter of fact, some gazelles live their entire lives without ever taking a drink of water!

One way a gazelle is able to obtain water without having to drink is from the food it eats. The consumption of plants provides the gazelle with a very good source of moisture. The gazelle is efficient about the water it takes in via its food; it eats plants early in the morning or during the night, when the heat is not so oppressive. During these cooler parts of the day, the water content of plants is at its highest.

The gazelle's sleek coat is especially designed to help keep it cool. Their light-colored coat reflects the sunlight, while their white underbelly deflects away the heat that radiates upward from the ground.

Even more incredible is how the gazelle is able to keep itself from overheating by maintaining a cool head. The gazelle has its own brain-cooling mechanism which prevents a buildup of heat from affecting its brain. This structure, known as the carotid rete, is a configuration of blood vessels in the brain that can keep the brain's temperature lower than the temperature of the rest of the body. Maintaining a cool brain is essential for the gazelle's survival, as it gives the gazelle the ability to continue to run without overheating when being chased by a predator. A predator like the cheetah, for example, must stop running when its body and brain temperature reaches 105°F, whereas the gazelle can keep running even as its body temperature rises above 109°F, as its head's temperature remains much cooler than that.

Warning! Tommies stamp their front feet to signal when they are disturbed.

Wacky Fact:
Tommies are easily identified by their tail motion. Their tails swing from side to side like a windshield wiper!

Torah Talk

Naftali is a messenger gazelle who delivers beautiful sayings (Bereishis 49:21).

This was said as a blessing to Naftali by his father Yaakov Avinu. Naftali was blessed with the special character traits of the gazelle: speed, charm, and grace. To better appreciate Naftali's *brachah*, let us first understand the *pasuk's* reference to "a messenger gazelle."

The Ramban brings down the *Yerushalmi* (*Shevi'is* 9:2) that says that in ancient times, rulers used to send gazelles as message carriers to one another. They trained their gazelles in the following manner: The gazelles that were born and raised in the south would be sent north, and the gazelles born and raised in the north would be sent south. Then, when a king from the north needed to send an important message to a king in the south (or vice versa), he would tie a note onto the horn of the gazelle and let it loose. The gazelle would run speedily, directly to its birthplace, and in this way the other king would receive the message directed to him. Thus, the gazelle would be the "messenger" of many "sayings."

According to the *Targumim*, the name "Naftali" is made up of the two words *"nofes lo,"* meaning "the honey which flows from him." The Abarbanel explains that this is a reference to the tribe of Naftali's reputation for eloquence.

As a result of his natural abilities and his graceful manners, the tribe of Naftali was predestined to be the bearer of good news. Naftali himself had the occasion to fulfill this role in his lifetime, as some sources state that it was Naftali who first announced to Yaakov that Yosef was still alive (see *Targum Yonasan, Bereishis* 49:21). Later, Naftali's descendants proved themselves to be excellent diplomats, thanks to their charm, cleverness, and skillful grace (see *Midrash Tanchuma, Vayechi* 13). Naftali became the messenger who did not fail in his mission; he was capable of carrying out the most difficult tasks, for his penetrating and inspired spirit and his disarming grace made him an exceptional bearer of news in the highest sense of the term.

Record Holder:

In the summer of 2007, a mega-herd of 250,000 Mongolian gazelles was sighted off the country's steppes. It was the largest gathering of gazelles ever recorded. Usually, there are "only" around 5,000 gazelles traveling this route.

INTERESTING FACTS & STATS

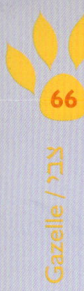

🐾 The Thomson's gazelle and the cheetah are the two fastest animals on land, with cheetahs able to attain higher speeds, but Thomson's gazelles able to outlast the cheetahs in long chases and able to make turns more speedily.

🐾 The male Grant's gazelle has larger horns in proportion to its body size than any other antelope.

🐾 The Dama gazelle is the largest type of gazelle. It is about forty-three inches high at the shoulder and weighs up to 190 pounds.

🐾 The Sand gazelle is not a great leaper; instead, it evades predators with its remarkable bursts of speed, sometimes reaching sixty miles an hour!

🐾 Female gazelles are called does.

🐾 Male gazelles claim their territory by marking them in an unmistakable way: placing a small amount of secretion from their scent glands (located beneath the eyes) onto a blade of grass within their boundaries.

🐾 The male gazelle will aggressively defend its territory against a rival male gazelle. When challenged, the two gazelles will clash horns, and the victor will claim the territory.

🐾 Female gazelles' horns tend to be shorter than the horns of male gazelles.

🐾 Female Thomson's gazelles are able to give birth twice a year.

🐾 The only relatively long-lasting relationship among gazelles is that of a mother and her most recent offspring.

Gazelle Trivia

1. **What is a male gazelle called?**
a. a bull b. a tom-tom c. a buck d. a doe

2. **A gazelle is considered to be a…**
a. type of deer b. type of cat c. type of dog d. type of antelope

3. **The name "gazelle" comes from an Arabic word that means…**
a. the fast one b. to be affectionate c. beautiful d. to be graceful

4. **What do you call a group of gazelles?**
a. a pack b. a herd c. a tribe d. a mob

5. **Which one of these statements is false?**
a. Male gazelles are born with horns. b. Gazelles have an exceptional sense of sight.
c. Gazelles have an excellent sense of smell. d. Gazelles have an exceptional sense of hearing.

ANSWERS: 1. c 2. d 3. b 4. b 5. a

Animal Crackers:
Q. Why did the gazelle wear a bell around its neck?
A. Because its horns were not working.

GIRAFFE גירפה

The Giraffe

The giraffe is a gorgeous and gentle creature that stands out among animals like a skyscraper among buildings. Typically, these fascinating animals roam the open grasslands in small herds. Thanks to their towering legs and long necks, giraffes are the world's tallest mammals. A male giraffe can grow to be eighteen feet tall.

Zoologists call the giraffe, *Giraffa Camelopardus*. In olden times, the giraffe was called a "camelopard," since it has spots like a leopard and walks a bit like a camel.

The long legs (six feet long) of the giraffe allow them to run as fast as 35 mph. Giraffes have a way of moving, or gait, in which both the front and back legs on one side move forward together, and then the other two legs on the other side move together. In this manner, the giraffe takes steps that are up to fifteen feet long. This is called *pacing*.

Predators — Beware!

The giraffe's height helps it keep a sharp lookout for predators across the wide expanse of the African savanna. Giraffes are so big that they do not need to hide from predators. Besides humans, they are hunted only by lions and crocodiles, and their speed helps them escape from these predators.

When they have to, giraffes will defend themselves with a deadly kick. With a single kick, a giraffe is able to knock off the head of a lion!

Big Pain in the Neck

The giraffe's six-foot-long neck weighs about 600 pounds. Just like humans, giraffes have only seven neck vertebrae; for giraffes, however, each one can be over ten inches long!

Male giraffes use their necks as weapons in combat, a behavior known as *necking*, which is used to establish dominance. There are two types of necking: *low intensity necking* and *high intensity necking*. In low intensity necking, the two giraffes rub and lean against each other. The giraffe that can hold itself more firmly wins the round. In high intensity necking, the two giraffes will spread their front legs and swing their necks at each other, attempting to land blows with their *ossicones* (hair-covered horns). Both male and female giraffes have these distinct hair-covered horns, although the female giraffe's horns are shorter than the male's and usually have hair on top of them, too.

Did You Know?

A giraffe has never been observed bathing.

THE CIRCLE OF LIFE

The Circle of Life

After a pregnancy of 14–15 months, the female giraffe gives birth, usually to a single baby. When a giraffe baby, called a *calf*, is born, it drops to the ground head first, about a six-foot drop!

The mother then grooms the newborn and helps it stand up. A newborn giraffe is about six feet tall and weighs 150 pounds. Within a few hours of birth, the calf is able to run around.

During the first week of its life, the mother carefully guards her calf. Young giraffes are very vulnerable and cannot defend themselves. For the first 2–3 weeks, the little giraffe spends most of its time hiding; its coat pattern provides good camouflage.

Mothers with calves will gather in nursery herds, moving or *browsing* (feeding on leaves and shoots) together. Mothers in such a group may sometimes leave their calves with one female while they forage and drink elsewhere. This is known as a *calving pool*.

Calves are at risk of predation, and a mother giraffe will stand over her calf and kick at an approaching predator. The bond a mother shares with her calf varies, while the adult males play almost no role in raising the young. The calf stays with its mother for 1½ years. Young giraffes become mature when they are four years old, and they are fully grown when they are six years old.

The average lifespan of the giraffe living in the wild is twenty-five years.

What's for Supper?

Giraffes spend between sixteen and twenty hours a day feeding. A giraffe may eat up to seventy-five pounds of food per day. It takes a lot of leaves to fuel such a large animal. Their favorite leaves are from the acacia tree. Acacia trees have long thorns that keep most animals from eating its leaves. But these thorns don't stop the giraffes! They simply use their eighteen-inch-long, strong tongue, which is coated with a thick layer of glue-like saliva to protect it, to pluck the tasty morsels from the branches.

Since the giraffe eats hundreds of pounds of leaves each week, it needs to travel miles to find enough food. If they need to, giraffes can go for several days without water, staying hydrated by the moisture of the leaves they eat.

Baby Fact:

A giraffe is one of the few animals born with horns.

Unique Traits

Unique Traits

A couple of unique traits of the giraffe that are not visible on the surface involve its heart and circulatory system.

A huge heart — as well as a very high blood pressure — is necessary in order to pump the giraffe's blood all the way up to its head. In fact, the giraffe has the largest heart in the animal kingdom, measuring two feet long and weighing about twenty-five pounds! The giraffe's special heart is able to produce the necessary high blood pressure for the giraffe.

But there's more to the wonder involved in this. When the giraffe drinks, its head is suddenly lowered seven feet below the level of its heart...and then, a few seconds later, it might be lifted back to nearly twenty feet into the air! These implications should be devastating. When its head is lowered that quickly, the giraffe should be stricken with a brain hemorrhage, the result of a huge rush of blood to its head. Conversely, the rapid raising of its head should drain all the blood from it, causing immediate unconsciousness. The fact that neither happens is the result of a specially designed valve which instantly and effectively controls the blood supply, to prevent the dramatic fluxes in pressure from knocking out its owner.

In its upper neck, a complex pressure-regulation system called the *rete mirabile* prevents excess blood flow to the brain when the giraffe lowers its head. When the blood supply begins to get full, the pressure is felt by the giraffe in time for it to right itself. Then the process works in reverse; the veins and valves partially close off, so blood doesn't flow back down into the giraffe's body too rapidly. Humans do not possess this special valve, as they have no need for it. The giraffe, however, could not survive without it.

The degree of sophistication found in the giraffe is further testimony of the work of a Master Designer, and publicizes the greatness of Hashem and the brilliance of His creations.

Fill 'Er Up!
A giraffe can go longer without water than a camel.

Wacky Fact:
A giraffe is able to clean its ears with its own tongue!

Torah Talk

The giraffe is a cud-chewing animal, and its feet have split hooves. Thus it possesses the two signs which make it a kosher animal. In fact, major Torah commentaries such as Rav Saadiah Gaon, Rabbeinu Yonah, and Radak identify the *zemer*, listed among the ten types of kosher animals, as the giraffe (*Devarim* 14:5).

A common misconception is that giraffe meat is not eaten because we don't know where on the neck to slaughter it (as prescribed by the *halachos* of *kashrus*). In fact, the giraffe can be slaughtered anywhere on its six-foot-long neck.

The Gemara (*Chullin* 27a, 45a) and the *Shulchan Aruch* (*Yoreh De'ah* 20:1–2) give precise parameters indicating the top and bottom of the neck that define the area within which *shechitah* may be performed. For a pigeon, this area is a few inches long; for a cow, over twelve inches; and for a giraffe, close to six feet!

The halachic basis for not eating giraffe is because in addition to needing the physical criteria for kosher animals to be met, the Torah may also require a continuous *mesorah* (tradition) of actually eating the specific animal in question.

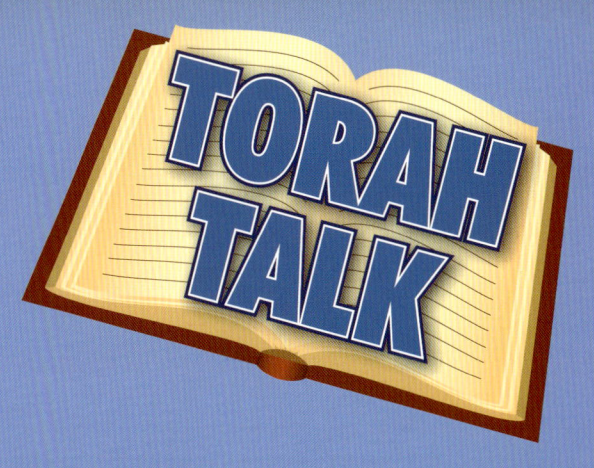

According to most halachic authorities, the need for such a *mesorah* is essential only in the case of birds; as Rabbi Yitzchak said, "Birds are eaten by *mesorah*" (*Chullin* 63b). As for animals, it appears from the Rambam that merely recognizing them as being kosher is enough (*Ma'achalos Asuros* 1:8). However, some Ashkenazic halachic authorities have ruled that animals also require a *mesorah*. Therefore, the giraffe, despite its signs of being a kosher animal and its long, *shecht*-able neck, would still not be permitted to be eaten without an uninterrupted *mesorah* of actually doing so.

That being said, there are also practical concerns that make the consumption of giraffe meat unreasonable. Slaughtering an animal of that size is no easy chore, particularly when you consider that one kick of a giraffe can kill a lion. Finally, even if methods of breeding and slaughtering could be found, the price of the meat alone would probably be exorbitant to most people.

Record Holder:
The tallest giraffe in the world was 19.3 feet tall and weighed approximately 4,400 pounds!

INTERESTING FACTS & STATS

🐾 Giraffes have a very strong odor that is intolerable to humans but protects the giraffes from many types of parasites.

🐾 Giraffes are ruminants. This means that they have more than one stomach. In fact, giraffes have four stomachs, the extra ones assisting with digesting food.

🐾 The average weight of a giraffe is 1,764 pounds.

🐾 Giraffes drink large quantities of water (ten gallons per day) when it is available, which enables them to live for long periods in arid areas.

🐾 Although giraffes seem very quiet, they in fact make a whole lot of different sounds — coughs, grunts, bleats, whistles — and even more sounds that are too low for us to hear.

🐾 Giraffes have the longest tail of any land mammal — up to eight feet long, including the tuft at the end.

🐾 Giraffes have beautiful spotted coats. While no two coats have exactly the same pattern, giraffes from the same area appear similar to each other.

🐾 The distinctive spots that cover a giraffe's fur act as a good camouflage in order to protect the giraffe from predators. When the giraffe stands in front of trees and bushes, the light and dark coloring of its fur blends in with the shadows and sunlight.

🐾 Giraffes have eighteen-inch-long, bluish-purple tongues which are tough and covered in bristly hair to help them with eating leaves from the thorny acacia trees.

Giraffe Trivia

1. What color eyes do giraffes have?

a. black b. light brown c. dark brown d. purple

2. What is a male giraffe called?

a. bull b. cow c. stallion d. geoffrey

3. How long does a giraffe sleep during a 24-hour time period?

a. It never sleeps. b. 1–2 hours c. 4–6 hours d. 8–10 hours

4. The giraffe got its name from the Arabic name 'Xirapha.' What does this name mean?

a. the gigantic one b. the smelly one c. the majestic one d. one who walks swiftly

5. Which one of these statements is false?

a. Giraffes are born with a horn. b. Giraffes have poor eyesight. c. Giraffes have four stomachs.

ANSWERS: 1. c 2. a 3. b 4. d 5. b

Animal Crackers:
Q. Why do giraffes have such long necks?
A. Because their feet smell.

GOAT עֵז

The Goat

There are many different species of goats in the world; however, the most common is the domestic goat, which is a subspecies of the wild goat. Goats have cloven hooves; long beards on their chins; short tails; and straight hair. In the winter, they grow woolly undercoats. Male goats and most female goats have two horns that grow on top of their heads, but the males' horns are much larger. The horns are made of keratin, like our fingernails, and they are permanent, growing throughout the goat's lifetime. The male goat, called a buck, uses its horns to defend itself from other dominant male goats and from its predators.

Goats also have very distinct eyes. The pupil in a goat's eye is rectangular in shape instead of being round. The width of the pupils allows the goat to see at a 330-degree angle, as opposed to humans, who see at around a 185-degree angle. Goats have excellent night vision, and they will often look for food at night.

Most goats live in herds of about five to twenty animals. Goats prefer being part of a herd and will generally live longer while living with a herd, as opposed to living on their own. Every herd has a "herd queen," who leads the herd while browsing (eating twigs and leaves). The herd queen is generally the oldest, and gets the privilege of eating first.

Goats are extremely curious and intelligent. They are also very coordinated and can climb and hold their balance in the most awkward places. They are typically found in more barren landscapes, and many species of goats even prefer mountainous and rocky terrains. The goats that inhabit the mountainous cliffs are amazingly agile. They are able to walk on small ledges and are very adept at jumping and running around on them, as well. Some goats can walk along a ledge not much wider than a tightrope, and some goats are even able to climb trees!

The most common domestic goat breeds are the Angora, Cashmere, French-Alpine, Nubian, Saanen, and Toggenburg. The Angora and the Cashmere are the main breeds of hair-producing goats.

Did You Know?

Mother goats do not forget the sound of their kids' voices, even a year after they have been weaned and separated from them.

THE CIRCLE OF LIFE

The Circle of Life

The female goat, called a doe, usually has a five-month pregnancy. Before her labor begins, the doe will leave her herd and find a quiet place where she can give birth and bond with her newborns. A newborn goat is called a *kid*. Does usually give birth to twins; however, it is not uncommon for it to give birth to triplets or even to a single kid. In rare cases, a doe can give birth to a litter of up to six kids.

The average weight of a newborn kid is six pounds. It is usually able to walk within one hour after birth. Some kids can even jump within their first day of life. Once the kid is strong enough to follow its mother, they will both join the herd. Although kids become independent at a very early age, they continue to nurse from their mother for up to six months. A kid usually remains with its mother for at least one year.

Goats mature in about a year, and can live for up to fifteen years.

What's for Supper?

Many people think that goats will eat almost anything, including tin cans and cardboard boxes. However, this is not accurate. Goats do not actually *consume* inedible material. Since goats are browsing animals, not grazers like cattle and sheep (grazers eat grass, while browsers eat twigs and leaves from trees and shrubs), and are naturally curious creatures, they will *nibble on* and *taste* just about anything that looks like plant matter, including cardboard and paper labels from tin cans. Then they can decide whether or not the substance is good to eat.

Goats are actually very particular about what they eat. They will not eat soiled food unless they are facing starvation. They prefer to eat the tips of trees, shrubs, and broad-leaved plants, although their diet also includes woody plants, weeds, and briars. Interestingly, goats have a higher nutrient requirement than larger ruminants.

Baby Fact:
When female goats give birth to babies, called kids, the process is called *kidding*.

Unique Traits

Unique Traits

Goats possess unique climbing capabilities, which enable them to live in some of the world's steepest and most treacherous habitats. They have an excellent sense of balance and enjoy jumping and climbing games with their herd mates. Wild goats, when climbing, can cling to the tiniest ledges of a cliff that is almost vertically straight, while domesticated goats enjoy climbing high walls and walking atop very narrow fences. Besides being very sure-footed, goats are also very agile and are expert leapers. They excel at making long, flying leaps from rock to rock, always landing with both front feet close together.

Markhors, a species of wild goats found in Central Asia, are particularly well known for their spectacular climbing abilities. These goats can be found living on the dangerous hillsides of the Himalayan Mountains, and are able to climb as high as 13,000 feet above sea level. Oddly enough, Markhors are also known for their aptitude at tree-climbing.

Tamri goats, found in western Morocco, are another species of goats that can climb trees. They are able to scale the steep and narrow trunks of Argan trees, and then balance themselves on the branches, in order to feed on the trees' leaves and nuts.

How can goats do all this?

The secret of the goat's sure-footedness is in the construction of its hoof. The goat's hoof has a hard outer shell and a soft, concave, inner footpad, which acts like a suction cup when weight is applied to it. The goat's hoof is also cloven, or split into two toes. This gives the goat a much better grip on the surfaces that it climbs, as each half of the hoof has its own hold on the area. The goat's cushion-like soles provide the goat with superb traction, allowing the animal to keep its balance as it moves across uneven or slippery grounds and dangerous terrains.

Another benefit of the soft cushioning inside the goat's hooves is that it acts as a shock absorber, enabling the goat's leg muscles to suffer no strain at all when the goat lands after a jump, even if it's on a hard surface.

Wacky Fact:

One of the rarer species of goats is the fainting goat. This goat literally freezes up; its legs go rigid, and the goat falls over. Soon after, it gets back up and continues browsing, until the "fainting" happens again.

Torah Talk

According to the midrash (*Midrash Rabbah, Vayikra* 27:9), the types of animals that qualify to be brought for *korbanos* were all chosen in the merit of the *Avos*. The ox was chosen in the merit of Avraham; the sheep was chosen in the merit of Yitzchak; and the goat was chosen in the merit of Yaakov.

The mishnah in *Pirkei Avos* (1:2) states: *The world stands on three principles — Torah, service of Hashem, and acts of kindness.* Hashem chose the ox to qualify for an offering because it represents the *chessed* of Avraham, as Avraham served tender calf to his guests, the three *malachim* who visited him. The sheep was chosen because it represents the sacrifice and service of Yitzchak, as it was a ram that was found in the thicket and offered as a sacrifice at *Akeidas Yitzchak*. But how does the goat represent Yaakov, the pillar of Torah?

The Torah states regarding Rivka Imeinu telling her son Yaakov to take the blessing from his father Yitzchak: *"So now, my son, heed my voice to that which I command you. Go now to the flock and fetch from there two good goats..."* (*Bereishis* 27:9-10). The goats that Yaakov brought to his father were the means through which he received the blessings from Yitzchak.

The above midrash explains that these goats brought about two positive results. Firstly, Yaakov received the blessings. Secondly, in the merit of these goats, Yaakov's descendants, the Jewish people, merited to be atoned each year through two other goats that were brought on Yom Kippur in the Beis Hamikdash, the *se'ir la'Hashem* and the *se'ir la'Azazel*.

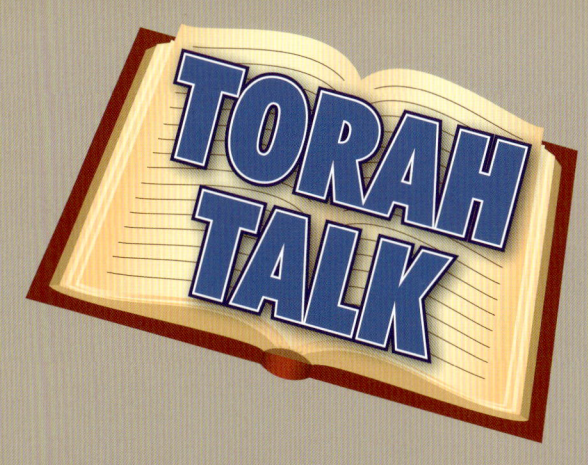

If Yaakov had not taken the blessings when his mother instructed him to do so, there would not have been a Jewish people. The blessings would have gone to Eisav instead, leaving Yaakov no material wealth with which to support and sustain himself. The ultimate fulfillment of the purpose of existence, which is the Jewish people accepting and fulfilling the Torah, was able to come about only through the goats that Yaakov had taken. Furthermore, it was only in the merit of the other two goats, the *se'ir la'Hashem* and the *se'ir la'Azazel* (which themselves were in the merit of the original two goats of Yaakov), that Klal Yisrael were forgiven each year, thus allowing for a continuation of their existence. Therefore, the goat represents the basis for all *kiddush Hashem* that will come about until the end of time.

Going Up!
The ibex, also known as a wild goat, can jump over six feet straight up from a standing position!

Record Holder:
The world record for the longest goat horns belongs to a goat named Uncle Sam.

INTERESTING FACTS & STATS

🐾 World-wide, more people drink goat's milk than cow's milk.

🐾 Goats live on every continent except Antarctica.

🐾 The goat has two musk glands behind the base of its horns. These musk glands are what cause the goat to have a bad smell.

🐾 Goats have no upper front teeth. Instead, they have a hard "gum pad."

🐾 The age of a goat can be determined by the size and condition of its lower front teeth.

🐾 Both female and male goats have beards, and some have wattles, which are the little dangly pieces of skin on the sides of their necks, just below their jaws.

🐾 A male goat is referred to as a "buck" or "billy goat." A female goat is known as a "doe" or a "nanny goat," and its offspring, up to the age of one, is called a "kid."

🐾 Goats' hair was used for making the curtains of the Mishkan in the desert (*Shemos* 26:7). The women spun the goats' hair right off the backs of the goats while the goats were still alive.

🐾 The cry of the goats is referred to as "bleating." Goats communicate with each other by bleating.

🐾 Goats will stand on their back legs to reach tree branches and shrubs.

Goat Trivia

1. A group of goats is referred to as a "herd," but it can also be called…
a. a bunch **b.** a school **c.** a trip **d.** a google

2. It is most common for a female goat to give birth to…
a. two kids **b.** three kids **c.** six kids **d.** eight kids

3. Which statement is false?
a. Female goats do not have beards. **b.** Male goats are called bucks.
c. All goats have rectangular shaped pupils. **d.** Goats are good swimmers.

4. Which goat has long and silky fleece, better known as mohair?
a. the Cashmere **b.** the Angora **c.** the Mohair **d.** the ibex

5. What type of goat is the mountain goat?
a. a domestic goat **b.** a wild goat **c.** an extinct goat **d.** It is not a true goat.

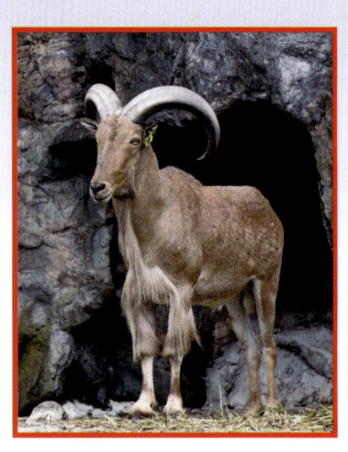

ANSWERS: 1. c 2. a 3. a 4. b 5. d

Animal Crackers:
Q. Why is it hard to have a serious conversation with a goat?
A. Because it's always kidding.

HIPPOPOTAMUS סוס-היאר

The Hippopotamus

Despite the hippopotamus's innocent-looking appearance, it is one of the most dangerous of all mammals. It is considered to be one of the most feared animals in Africa (as well as one of the largest), as both males and females are known to be incredibly aggressive.

The hippo has an enormous, gray, barrel-shaped body that can measure up to fifteen feet in length. A full-grown male hippo can weigh up to 7,000 pounds, while a full-grown female hippo can weigh up to 4,000 pounds. Even though the hippo looks like it would be slow on land due to its size and its short and stubby legs, it is actually capable of running at quite remarkable speeds — up to 30 mph!

The hippo has a huge head, which makes up around a third of its total body weight. Its vast mouth can open up to 150 degrees, revealing the hippo's enormous jaws, one of its most distinctive features, and the two long canine teeth (tusks) contained within them. These tusks can grow to be up to twenty inches long and are used mainly for defense or fighting with other hippos.

The hippo tends to live in small herds containing between ten and thirty animals. They have been seen, however, in groups as large as two hundred animals. The herd will usually live beside a river, where the territory can stretch for approximately 250 miles. The herd is led by the dominant male, who will fiercely guard its stretch of the riverbank from both intruders and rival males, threatening them by opening its mouth to expose its impressive tusks. It may also yawn, scoop water with its mouth, shake its head, rear up, roar, grunt, and make a loud wheezing sound, all of which are threat displays. If this fails, it will fight its enemy, and deadly injuries will often occur.

Lions, crocodiles, and hyenas are the most common predators of the hippo. However, due to its size and strength, an adult hippo is very difficult for predators to kill and is rarely attacked. Instead, predators will often prey on a sick hippo or on a baby hippo.

Female hippos and their young congregate in large groups in order to intimidate and ward off their predators.

Did You Know?
The only two land animals that are larger than the hippopotamus are the elephant and the white rhinoceros.

THE CIRCLE OF LIFE

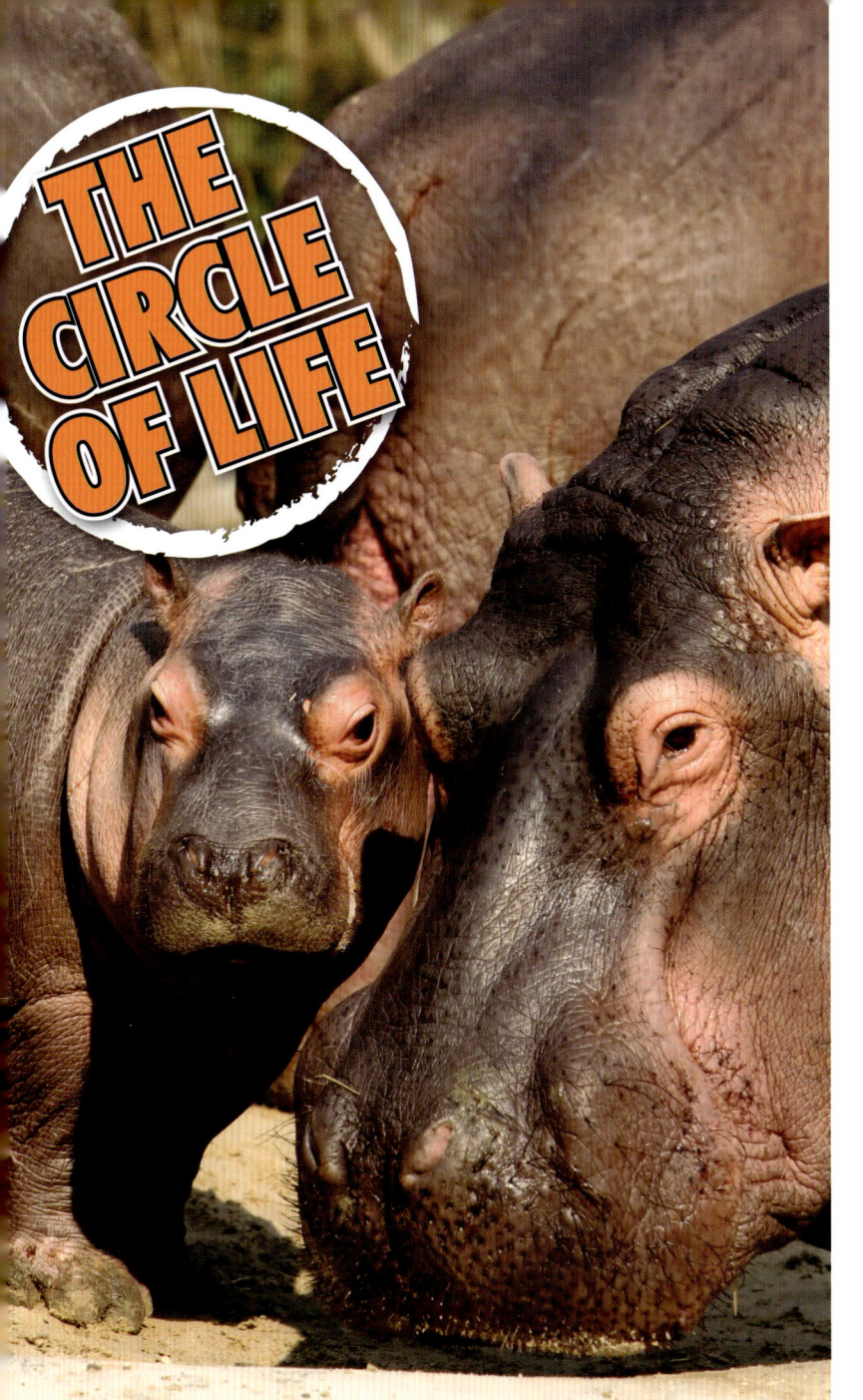

The Circle of Life

The female hippopotamus, called a cow, usually has only one offspring every two years. It will generally give birth to a single baby hippo, called a calf, although twins may sometimes occur. The hippo will often give birth in the water, where most of its activities are done; however, some calves are born on land, as well. The baby calf weighs nearly one hundred pounds at birth and is able to swim almost from the moment it is born. When a baby hippo is born underwater, the mother needs to push it to the surface to breathe, since the newborn is only able to hold its breath for about forty seconds at a time.

While hippos are known to be extremely aggressive, the mother hippo is a very caring parent. The mother hippo stays with its newborn calf for several days, without leaving it even to go eat. It will nurse its calf for eight months, or even up to a year if food is scarce. The mother hippo is also very protective of its young. As a safety precaution, the calf will often ride on its mother's back when they are in water that may be too deep for the calf.

The hippo calf is fully weaned by the time it is eighteen months old, but tends to remain with its mother until it is fully grown, often not leaving her until it is seven or eight years old.

The average lifespan of a hippo living in the wild is forty-five years.

What's for Supper?

In the evenings, when the sun goes down and the temperature is cooler, the hippo will come out of the water and search for food. Despite the hippo's enormously long and sharp teeth, it is an herbivorous animal. Its diet mainly consists of short grasses. Oddly enough, the hippo doesn't use its large canines for eating, as it doesn't chew with its teeth, but with its lips. Despite its large size, the hippo only eats around one hundred pounds of food a night (about one to one and a half percent of its body weight), as it uses very little energy while relaxing in the water for most of the day. The hippo can store two days' worth of grass in its stomach, and can go up to three weeks without eating, if necessary.

Baby Fact:
A baby hippo is able to nurse underwater! At first, it will nurse for thirty-five seconds and then come up for air, but in time it will build the capacity to nurse for complete two-minute intervals.

Unique Traits

The hippopotamus is a semi-aquatic animal that spends up to eighteen hours a day in the water. The hippo was created with a very unique set of character traits in order for it to be well-suited for such a life. Since it lives in a habitat that is extremely hot during the day, it is necessary for the hippo to live in the water to escape the immense heat from the sun.

The hippo's eyes, ears, and nostrils are located at the top of its head, which means that when the hippo is immersed in the water, it is still able to see, hear, and breathe. Incredibly, when the hippo needs to submerge itself completely underwater, it is still able to see. This is because its eyes are equipped with a special, clear membrane that protects them while still allowing them to see when underwater!

Another amazing fact about the hippo is that its ears and nostrils automatically shut upon contact with water. In addition, it is able to breathe underwater for five to six minutes, and can even remain underwater for as long as thirty minutes, if necessary.

The hippo has four webbed toes on each foot, enabling it to travel along the slippery banks of the river. Yet, despite its love of the water, the hippo is not a very good swimmer. Its body is too dense to float, so instead it is only able to walk along the riverbed or push off from the bottom of the river.

The hippo often likes to relax in the river in order to absorb enough water to keep its skin moisturized. The hippo has very unique skin that needs to be kept wet for a good part of the day to prevent dehydration. Its skin does not have true sweat glands; instead, it secretes a thick, oily, red substance from its pores known as "blood sweat," because it looks like the hippo is sweating blood. This creates a layer of mucus that protects the skin from sunburn and keeps it moist. Additionally, it is thought that this mucus may also work as an antiseptic which prevents infections, since most of the rivers that hippos live in are dirty and bacteria-filled, yet the hippo's wounds do not get infected.

What's in a name? The word "hippopotamus" comes from two Greek words that mean "river horse."

Wacky Fact: The milk of a hippopotamus is bright pink!

Torah Talk

Did you know that at first glance, the hippo seems to have the signs of a kosher animal? Yet, as you likely already know, there is no such thing as kosher hippo meat; the hippopotamus is not a kosher animal at all. Why not?

Before answering this question, we need to review the kosher laws pertaining to meat.

Most meat is not kosher. That is because most animals are not kosher. A kosher animal has two distinguishing characteristics: it has split hooves and it chews its cud. The majority of animals do not have both features.

What are split hooves? Look at the animal's foot. Does it have a paw or a hoof? If it has a paw, it is not kosher. If it has a hoof, is the hoof a big block — like a horse's hoof? Or is it split in two — like a cow's hoof? If it is split, you know it has one distinguishing characteristic.

The other necessary characteristic is cud chewing. A kosher animal has multiple chambers in its stomach. The animal eats grass or grain, chews it, swallows it, brings it back to its mouth to chew again, and then swallows it into a different chamber of the stomach. An animal that does this is called a ruminant — a cud-chewing animal.

Where does the hippopotamus come in with all this?

The hippopotamus has a multiple-chambered stomach, and it seems like it has split hooves, too. Based on this, one might conclude that a hippo is indeed a kosher animal. However, upon closer inspection, one will see the differences between kosher animals and the hippo.

Firstly, without exception, every animal with a four-chambered stomach is a ruminant. There are those who dispute this and assert that the hippo is an example of an animal that has four chambers and is not a ruminant, but this is erroneous. In fact, the hippo has a three-chambered stomach! Therefore, the hippo is not considered to be a cud-chewing animal.

Regarding its feet, many people think that hippos have split hooves; however, the hippo does not have a true cloven hoof. The feet of the hippo terminate in four developed toes that are covered and connected by thick skin. This produces a web effect that is ideal for swimming.

So, in conclusion, a hippopotamus is much like most non-kosher animals which do not chew their cud and do not have split hooves.

Record Holder:
The oldest hippo ever was called Tanga; she lived in Munich, Germany, and died in 1995 at the age of 61.

INTERESTING FACTS & STATS

- When rival male hippos confront each other, they stand nose to nose with their mouths open as wide as possible. This is called gaping, and it is a way for the hippos to size each other up. Generally, the smaller and weaker hippo will back away without being pursued by the larger hippo.

- Since male hippos are a threat to the calves, female hippos attack males that come too close. If a female begins to attack, the male hippo will lie down and act submissively to show that he means no harm.

- A hippo's hide alone can weigh as much as half a ton.

- Hippos have a complex form of communication that relies on grunts and bellows. Some zoologists think that hippos may even use echolocation as a means of communication.

- A male hippo will attack a calf in the water, but not on land.

- Hippos may travel up to three miles during the night to find a suitable place to graze.

- Although a hippo might look overweight and out of shape, it can easily outrun a human.

- Hippos are hostile toward crocodiles, which many times live in the same rivers as them and prey upon the baby hippos. Hippos have also been known to attack humans, and even boats.

- The hippo is one of the noisiest animals in Africa. Its vocalizations have been measured at 115 decibels.

Hippo Trivia

1. A group of hippos can be called several different names. Which one of the following names does **NOT** refer to a group of hippos?
a. herd b. pod c. bloat d. streak

2. The hippo's yawn is a sign of…
a. boredom b. sleepiness c. hunger d. threatening danger

3. According to a recent DNA study, the hippo is closely related to…
a. the whale b. the pig c. the horse d. the rhino

4. A male hippo is called a…
a. sow b. stallion c. bull d. hipster

5. How long is the tail of the hippo?
a. two to six inches b. eight to twelve inches c. sixteen inches d. twenty to twenty-four inches

ANSWERS: 1. d 2. d 3. a 4. c 5. d

Animal Crackers:
Q. What do you call an insincere hippo?
A. A hippocrite

HORSE סוס

Spotted!

In 1805, explorers Louis and Clark discovered the Appaloosa breed, which was possessed by the Nez-Perce Indian tribe.

Fast Fact:
Many Appaloosa horses are used for racing and have been Kentucky Derby winners.

The Appaloosa

The Horse

There is only one species of the domestic horse; however, there are around four hundred different breeds of it throughout the world. Most horses are domestic. The breeds that live in the wild are called feral horses. Although a feral horse is a free-roaming horse, it is not a true 'wild' animal since its ancestors were originally domesticated. (For example, the Mustang, a type of feral horse, is a descendant of horses brought by Europeans more than four hundred years ago.)

The four breeds discussed in this section are most noted for their speed (American quarter horse), strength (Belgian draft horse), beautiful colored coats (Appaloosa), and the distinction of being the only true wild horse in existence (Przewalski's horse).

The Appaloosa

The Appaloosa is best known for its colorful, spotted coat pattern. The coat color of an Appaloosa is usually a combination of a base color with an overlaid spotting pattern. Coat colors include: "leopard" — large, dark spots covering a white body; "snowflake" — a dark body with light spots; "marble" — a light coat covered in small, dark speckles; "frost" — a dark coat covered in small, light speckles; and "blanket" — where the coat is white just on the hips. However, some Appaloosas are "solid," meaning that they do not have any coat pattern.

Although there are various body types within the breed, most share the same physical characteristics, such as mottled black and white skin, vertically striped hooves, white sclerae which encircle the irises of the eyes, and a short mane and tail. Appaloosas are very versatile — they can be used for racing, riding, or ranch activities — and are famous for their intelligence, speed, stamina, and endurance.

The Appaloosa was originally bred in the inland northwest of America by the Nez-Perce Indians, where they were developed into fast and agile horses. The Nez-Perce, however, never referred to their horses as "Appaloosas." The name Appaloosa comes either from the Palouse River, near which the horses were plentiful, or from the Palouse tribe, whose village was on the Palouse River. Settlers first referred to the horses as "A Palouse horse," which was soon shortened to "Appalousey." The name Appaloosa was made official in 1938.

Today, the Appaloosa is one of the most popular breeds in the United States.

Did You Know?

On March 25, 1975, the Appaloosa was named Idaho's state horse.

Importer of the Year: The first Belgian draft horse was imported to America by Dr. A. G. Van Hoorebeck of Illinois in 1866.

Baby Fact: Foals, or newborn horses, can stand up and take steps within an hour after they are born!

The Belgian Draft Horse

A draft horse is a large, strong horse used for heavy labor such as pulling loads, plowing, and various farming tasks. Although there are a variety of draft horses, they all share many of the same physical traits. They are big, powerful, and muscularly built. Yet despite their enormous size and strength, they are very patient and docile. Popular breeds include the Clydsdale, Percheron, and the Shire. Although there are a number of different breeds of draft horses in the world, the Belgian draft horse is by far the most popular today.

The Belgian draft horse, as its name suggests, originated from Belgium. Belgium lies in the area of western Europe that was known for its large black horses called Flemish horses. They were referred to as the "Great Horses" by medieval writers. These horses are the ones that carried armored knights into battle. The Belgian draft horse is said to be directly descended from the "Great Horse." Most other draft breeds are descended either from the Belgian, or from the Belgian's direct ancestor, the "Great Horse," as well.

The Belgian draft horses are the widest, lowest, and most compact of the draft breeds. They are five to six feet high, and weigh from 1,900 to 2,200 pounds. Their colors normally are a shade of light chestnut called sorrel, or else red roan, dun, brown, or gray.

The Belgian draft horse is known for its kind temperament, and is easy to handle. These horses are still used for all sorts of draft work, including plowing, logging, and pulling heavy loads. However, due to modern technological innovations in the farming industry, their importance to farmers has declined and they are no longer used for the sole purpose of farming. Now, they are also used as show horses, gaming horses, and even as trail-riding horses.

Today, there are more Belgian draft horses than any other breed of draft horse in the United States. They outnumber the total of all the other breeds combined.

Record Holder:
The world's largest Belgian draft horse was named Brooklyn Supreme. It weighed 3,200 pounds and was six feet four inches tall.

MIA: Przewalski's horses have not been seen in the wild since 1968!

Wacky Fact: Przewalski's horses have 66 chromosomes, whereas domestic horses carry only 64!

The Przewalski's Horse

Przewalski's horse, a rare and endangered animal, is the last surviving subspecies of the wild horse. As noted before, most "wild" horses today, such as the Australian Brumby and the American Mustang, are actually descended from domesticated animals that escaped and went to live in the wild. Przewalski's horse, however, has never been successfully domesticated and is considered the only true wild horse.

Przewalski's horses are named after Russian explorer and geographer Nikolai Przewalski, who first studied them. They were originally discovered along the border between Mongolia and China in 1879, but were not scientifically described until 1881, when Nikolai Przewalski found a skull and hide of this rarely seen horse and introduced them to researchers at a museum in St. Petersburg.

Przewalski's horses have short, muscular bodies and are smaller than most domesticated horses. They weigh about 440 to 750 pounds and stand forty-eight to fifty-six inches tall. They have large, rectangular heads with dark manes that are erect and spiky. Their coats vary in colors, from beige to reddish-brown, but they all have light bellies, darker backs, and white markings on their muzzles. Their coats are short during the summer, but grow long and thick in the winter, complete with long beards and neck hair.

Like their relatives, Przewalski's horses are grazers and mainly eat grass. When thirsty, they will use their sharp hooves to dig for water in the ground.

Excessive hunting by people, as well as losing their grazing and watering sites to domestic animals like cattle and sheep, decreased their population dramatically. In the late nineteenth and early twentieth centuries, several wild Przewalski's horses were caught and bred in captivity. Thirteen of those horses are the original ancestors of today's captive population. Today, there are about twelve hundred Przewalski's horses living in private preserves, zoos, and protected areas in Mongolia.

Did You Know?

Przewalski's horses became extinct in the wild around 1970. In 1994, some of these horses being held in captivity were reintroduced to the wild in a reserve in Mongolia.

Whoa! It is said that the quarter horse is able to "stop on a dime" from a full gallop!

Fast Fact: The American quarter horse can run a quarter of a mile in less than twenty-one seconds, starting from a flat-footed standstill!

American Quarter Horse

The quarter horse is considered a light horse breed. Light horse breeds generally weigh less than fifteen hundred pounds. They are typically used for pleasure riding. Due to their speed and agility, many are also used on the racetrack, in rodeo shows, and for farm work.

The American quarter horse excels at running short distances. Clocked at speeds of up to 55 mph, it is even faster than the Thoroughbred (a breed best known for its use in horse racing) over short distances. Due to its dominance over other breeds of horses in races on the quarter-mile track, it has earned the name "the quarter miler," or "quarter horse."

The American quarter horse, for all of its one-thousand plus pounds, is fairly short (sixty to sixty-four inches tall) and compact. It is extremely muscular, especially in its hindquarters, which is the source of the horse's great power for running. It has a short head and a very muscular neck.

The quarter horse was by far the most popular cattle horse in the early West. The origins of the quarter horse began in Virginia in the early seventeenth century, when these horses were given to American settlers by the Chickasaw Indians. The early quarter horses, aside from being used as riding horses, were ideal for the many tasks required for colonial life: hauling goods, pulling carriages, and farm work.

When the settlers began traveling west in the 1800s, the quarter horse breed was the horse of choice, being used for cattle round-up due to its speed and agility. It quickly became the horse that cowboys counted on for their daily chores on the ranch. As with most activities undertaken by the early cowboys, the duties performed by them and their horses quickly turned into competition, setting the stage for rodeo riding, calf roping, team roping, and barrel racing. Today, quarter horses are still in demand for these events.

The American quarter horse still remains the most popular breed in the United States today.

Did You Know?
When foals are born, their legs are almost the same length as they are when they are fully grown.

Unique Traits

Throughout history, the horse has always been the most influential animal in battle.

For thousands of years, ever since the horse was first domesticated, humans have used them in battle and warfare. There are records from China and Egypt that document the standard of horses pulling chariots into battle. The ancient Assyrians also recorded how they rode on horseback into battle, a concept later known as cavalry. In America, the conquistadores used horses to defeat the Aztecs and the Incas. By World War I and especially World War II, the cavalry became phased out for better technologies and advanced warfare. The Polish cavalry, however, remained active and attempted to defend itself several times against the Germans while on horseback.

None of this should be surprising, however, when you think into the nature of the horse.

Before Chava was created, Hashem brought every animal to Adam Harishon and asked him to choose a suitable name for it. Adam was blessed with the ability to give a name that accurately reflected the true essence and characteristic of each animal. When Adam first saw a horse, he named it "*sus*," because he recognized its gaiety and exultation in battle — "*sus*" is related to the word "*sas*," meaning "to exult." (See *Midrash Aggadah* 2:19, R' Bacheye.)

The most vivid portrayal of the horse, represented as a fighting animal, is found in Iyov: *The glory of his snorting is terrible. He paws with force, he runs with vigor, charging into battle. He scoffs at fear; he cannot be frightened; he does not recoil from the sword* (Iyov 39:20–22).

Nowadays, the horse continues to be used in law enforcement both in the United States and Europe. Mounted units are deployed in major cities throughout the world.

Wacky Fact:
Horses generally dislike the smell of pigs.

Torah Talk

A store owner once came to Reb Meir Premishlaner to complain about a new competitor who had recently opened a store across the street from him. The store owner feared that his livelihood was now in grave danger.

Reb Meir asked the store owner if he had ever noticed the mannerisms of a horse while it drinks water from a lake. While the horse drinks, Reb Meir explained, it stomps its feet on the ground. The reason for this is because it sees its reflection in the water and thinks there's another horse drinking by the lake as well. It stomps its feet to try to frighten this "other" horse away, for it is fearful that there will not be enough water for both of them. Of course, the horse is actually only stomping at itself — and there is certainly plenty of water.

"You," Reb Meir told the store owner, "are behaving just like the horse by the lake, since it, too, was afraid of an imaginary foe. Hashem has enough *parnassah* for everyone, and no one can touch or take anything that rightfully belongs to another person."

The person who has faith in Hashem, believing that everything comes from Him, understands that there is no reason to be jealous of anyone else. What one person possesses has no bearing whatsoever on what another person lacks.

Hashem provides each person with his or her specific needs. If a person has a desire to purchase a specific automobile and it was sold to someone else before he had a chance to purchase it, then it wasn't meant for him. If a person arrived late to a huge clearance sale and still found the item that he needed, it wasn't "good luck" or "perfect timing"; it simply meant that he was supposed to have that item.

There is no reason to be envious of someone else and behave like a horse — unless, of course, you are a horse!

Record Holder:
The oldest horse ever was called Old Billy. It was born in Woolston, Lancashire in 1760, and was sixty-two years old when it died on November 27, 1822.

INTERESTING FACTS & STATS

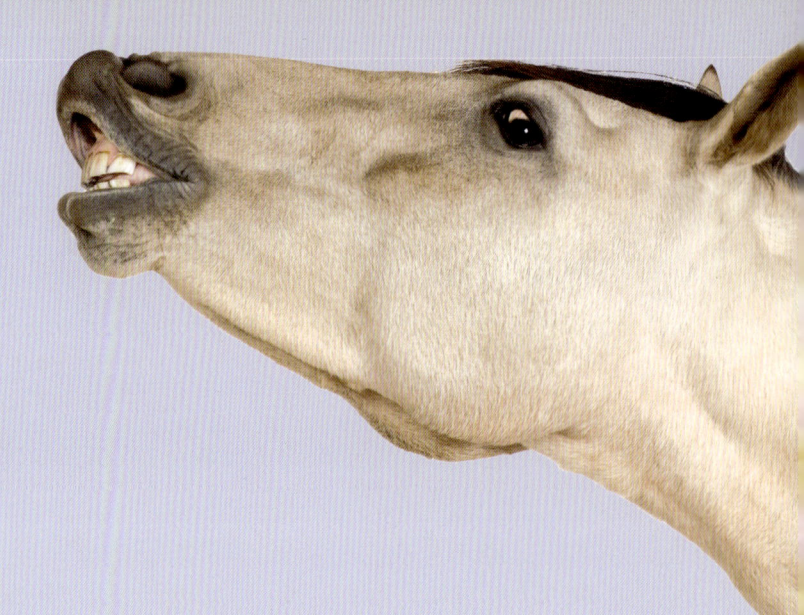

🐾 Male horses are called stallions, females are called mares, and babies are called foals.

🐾 A young female horse is called a filly, and a young male horse is called a colt.

🐾 You measure a horse's height in hands. Each hand equals four inches. If you say a horse is 15.2 hands high, the 2 stands for two fingers. Thus, a horse described as "15.2 h" is 15 hands plus 2 inches, for a total of 62 inches in height.

🐾 You can tell how old a horse is by its teeth — by how many teeth it has and by how the teeth change and wear down!

🐾 Any marking on a horse's forehead is called a star, even if it is not shaped like a star.

🐾 Horses are mammals in the same family as zebras, mules, and donkeys.

🐾 Horses can communicate how they are feeling by their facial expressions. They use their ears, nostrils, and eyes to show their moods. Beware of a horse that has flared nostrils and its ears back. That means it might attack!

🐾 Female horses give birth to one foal a year.

🐾 Horses can sleep lying down or standing up.

🐾 There is a breed of horses from Russia called Akhal-Teke. These horses can go for days without food or water.

Horse Trivia

1. Which horse is also known as the Asian wild horse?
a. Appaloosa b. Przewalski's horse c. Pinto d. Mustang

2. A horse typically sleeps for how many hours a day?
a. They don't sleep. b. less than one hour c. one to two hours d. two-and-a-half to three hours

3. How long do horses lie down during the day?
a. They don't lie down. b. less than one hour c. one to two hours d. two-and-a-half to three hours

4. Which horse has striped hooves?
a. Appaloosa b. Przewalski's horse c. Belgian draft horse d. quarter horse

5. A horse can move in several ways. Which of the following is not one of the ways?
a. walk b. canter c. sprint d. gallop

ANSWERS: 1. b 2. d 3. b 4. a 5. c

Animal Crackers:
Q. Why was the racehorse named 'Bad News'?
A. Because bad news travels fast!

KANGAROO קנגורו

The Kangaroo

The kangaroo is a marsupial. Marsupials are an infraclass of mammals living primarily in the Southern Hemisphere. A distinctive characteristic common to most species of marsupials is that the young are carried in an abdominal pouch called a marsupium. Kangaroos are the largest of all marsupials, standing over six feet tall.

Kangaroos have powerful hind legs; a long tail; large feet; and small front legs. It's difficult for them to move those huge back legs and feet one at a time. So, instead of running on all fours the way most animals do, they travel by hopping around on both legs. Their large feet enable them to leap thirty feet in a single bound. They use their strong tails for balance while jumping, and can travel as fast as 40 mph.

Kangaroos have small, furry heads; long ears; and pointed snouts. They can turn their ears from front to back to follow sounds. This helps them listen for danger. When a kangaroo senses danger, it alerts others by thumping its feet loudly on the ground. Kangaroos can also communicate with each other by grunting, coughing, or hissing.

Male kangaroos are powerfully built. When fighting with other kangaroos, they will lean back on their sturdy tail and "box" each other with their strong hind legs. They will also bite and use their sharp claws while fighting predators.

A group of kangaroos is called a *mob*, but can also be called a *herd* or a *troop*. There are more than fifty species of kangaroos. The two main types are red kangaroos and gray kangaroos.

Red kangaroos live in Australia's deserts and open grasslands. They normally move in mobs ranging from a few dozen to several hundred animals. They are the largest of all kangaroos, weighing up to two hundred pounds. The males have bright coats of thick, red fur. Female red kangaroos are smaller, lighter, and faster than males. They have a blue-hued coat, so many Australians call them "blue fliers."

Gray kangaroos roam the forests of Australia and Tasmania and live among the trees, though they do go to the open grasslands for grazing. They live in small groups but may congregate in large numbers when feeding. The Eastern Gray kangaroo is the best-known species of the kangaroo family. It has gray fur and weighs 145 pounds as an adult. The Western Gray Kangaroo is slightly smaller and lighter, weighing only 110 pounds.

Did You Know?
The kangaroo's family name, *Macropodidae*, means "big feet."

THE CIRCLE OF LIFE

The Circle of Life

Female kangaroos usually give birth to one baby at a time. The newborn kangaroo, called a *joey*, doesn't look like a kangaroo. At birth, the joey is the size of a coffee bean, and its hind legs are mere stumps. It has no fur, and it cannot see or hear. The joey immediately climbs into its mother's pouch and remains there for at least two months. Safe in that place, the newborn grows rapidly. The joey could not survive outside the pouch.

When the joey is three to four months old, it will leave the pouch for short trips and to eat small shrubs and grass. When it senses danger, it will quickly dive headfirst back into its mother's pouch. As a joey grows, its head and feet can be seen hanging out of the pouch.

At ten months of age, the joey is mature enough to leave the pouch for good. However, even after the joey has "moved out," it will continue to drink milk from the pouch until it is fully grown. By the time it is fully weaned, the young kangaroo may be over twelve months old. Kangaroos usually stay close to their mothers until they are two or three years old.

Amazingly, the mother kangaroo is permanently pregnant, except on the day she gives birth. The new embryo, however, is "frozen," with its development delayed, until the current joey occupying the mother's pouch is ready to leave. This ability of the mother kangaroo to delay its embryo's development is known as diapause, and it can also occur in times of drought and when food is scarce.

The red kangaroo can live up to twenty-three years in the wild, while the gray kangaroo can live there only up to ten years.

What's for Supper?

Kangaroos are herbivores, eating a wide range of plant matter. They use their small front legs to pull leaves from small plants and to dig into the ground for water. The gray kangaroo is mainly a grazer, eating a wide variety of grasses, whereas the red kangaroo includes significant amounts of shrubs in its diet.

Kangaroos' specialized teeth enable them to graze efficiently. Their incisors are able to crop grass close to the ground, and their molars chop and grind the grass. Kangaroo species that are adapted to the drier areas need very little water, while the red kangaroos can go without water altogether if there is fresh green grass available.

Baby Fact:
A female kangaroo can produce two different types of milk to feed two different babies at the same time: a joey that has left her pouch but is still nursing, and a newborn!

UNIQUE TRAITS

Unique Traits

A full kangaroo is an impressive animal. It can stand taller than a man, and commonly weighs two hundred pounds. Its huge hind legs are powerful enough to send it sailing over a fence, while a twitch of its tail can break a man's leg like a matchstick. Yet this same kangaroo begins life as a baby so tiny that it and two more like it could be held in a teaspoon!

A newborn kangaroo is less than an inch long! Its body is transparent and its eyes, ears, and hind legs are not at all developed. The only part of the baby that is fully developed is its little hands — and for good reason. Gripping its mother's fur, the young joey immediately makes the perilous climb to its mother's protective pouch, usually making the journey entirely on its own. No one has taught it the way, it cannot see where it is going, and it is nothing more than a tiny, underdeveloped embryo — but wondrously enough, this journey is always completed.

Once it has arrived at the pouch, further wonders follow. The joey takes hold of a milk gland, and hangs on with an inseparable grip. At first, the baby is not even sufficiently developed to suck for itself. For this reason, the mother kangaroo has been provided with special muscles with which she herself pumps milk to her baby.

The little kangaroo is attached by its mouth to the mother's milk gland for a few months straight. During that time, it cannot breathe through its mouth. How does it breathe? The answer is that while the young joey is nursing, an elongated part of its larynx connects with the back part of its nasal passages, so that air passes directly into its lungs from its nose. Thus, it can keep up its milk-drinking all the while and never choke.

Don't Worry – Be Hoppy! Hopping is more energy-efficient than running. The faster the kangaroos hop, the less energy they use!

Wacky Fact: Kangaroos cannot walk backward!

TORAH TALK

Torah Talk

Unlike other animals, the kangaroo is not found in various countries throughout the world. Kangaroos live exclusively in Australia, with some of the smaller species living in New Guinea. Nevertheless, the kangaroo was one of the wild animals included in the plague of *arov* (wild animals) that Hashem brought upon the Egyptians.

Animals tend to only eat produce that grows in their own habitats. As such, one could think that the trip to Egypt, so far from the kangaroo's home environment, would be a hardship for the kangaroos. However, this was not the case, as is evident from a *pasuk* in *Shemos*:

If you do not send out My people, I will send among you and among your servants and among your people and in your houses wild animals; the houses of Egypt will be full of the wild animals, as well as the ground they are on (Shemos 8:17).

There is something puzzling about this verse, which is describing the plague of *arov*. Having already said that the plague would affect all the houses and people of Egypt, why does the Torah have to say that it would also affect "the ground they are on"?

The Rogatchover Gaon offered the following possible answer: There is a halachah that states that we may not store the produce of the *shemittah* year in our homes after a particular kind of food is no longer available naturally to the *animals of the field* (Shemos 23:11). The mishnah in *Shevi'is* (9:2) states that for purposes of determining when the season for a particular food is over, Eretz Yisrael is divided into three sectors: Yehudah, the Galil, and Ever L'Yarden. When a food becomes depleted in one of these areas, that food (if it grew during the *shemittah* year) may no longer be consumed in that area, as it is no longer available to the *animals of the field* there, although it may still be found in the fields in other areas.

The Gemara in *Pesachim* (52b) asks why a food is considered to be unavailable to the *animals of the field*, simply when that food no longer grows in the fields where those animals live; can't those animals still eat the produce of the neighboring areas? The Gemara answers that in fact, the animals of one area would *not* eat produce from the neighboring lands, because "the animals of Yehudah do not eat the produce of the Galil."

Based on this Gemara, we can understand the deeper meaning of the phrase *the ground they are on*. Not only did the wild beasts wander from far and wide to torment the Egyptians, but also, soil from their natural habitats was miraculously supplied along with them. This way, even animals from far-off places, like the kangaroo, would be able to eat produce grown from the soil of their own natural habitats.

Record Holder:

The fastest recorded speed of any kangaroo was 40 mph. This record was set by a large, female Eastern Gray kangaroo.

INTERESTING FACTS & STATS

- Kangaroos can be tamed and trained to box playfully with humans.
- On land, kangaroos can't move their hind legs independently, only together. When they swim, however, they are able to kick each leg independently.
- Kangaroos are the only large animals that move by hopping.
- A single kangaroo kick can cause great harm to an enemy.
- The pouches of all kangaroos open forward, as opposed to the pouches of koalas or wombats that open backward.
- Kangaroos have very good eyesight, hearing, and sense of smell.
- Kangaroos are good swimmers, and will sometimes escape a threat by going into water if there is a lake or river nearby.
- Kangaroos do not sweat, so in the heat, they lick their front paws and rub the moisture onto their chests in order to cool down.
- The kangaroo is a national symbol of Australia.
- "Roos" is the slang term used for kangaroos.
- A male kangaroo is called a boomer, buck, or jack.
- A female kangaroo is called a flyer, doe, or jill.

Kangaroo Trivia

1. The kangaroo uses its tail for…
a. fighting b. swimming c. keeping cool d. balancing

2. Which type of kangaroo is the largest living marsupial?
a. Eastern Gray kangaroo b. giant kangaroo c. Western kangaroo d. red kangaroo

3. How many toes does a kangaroo have on its hind foot?
a. two b. three c. four d. five

4. Which statement is false?
a. Roos are good swimmers. b. Kangaroos have good hearing. c. Male kangaroos have pouches. d. A male kangaroo can be called "old man."

5. What is the color of the female red kangaroo?
a. red b. mostly blue c. mostly silver d. red and black

ANSWERS: 1. d 2. d 3. c 4. c 5. b

Animal Crackers:
Q. Can a red kangaroo jump higher than a ten-story building?
A. Of course it can — buildings can't jump!

LEOPARD נמר

The Leopard

Leopards are graceful and powerful big cats, closely related to lions, tigers, and jaguars. They live in sub-Saharan Africa, northeast Africa, Central Asia, India, and China. However, many of their populations are endangered, especially outside of Africa.

Most leopards are light colored with distinctive dark spots that are called rosettes, because they resemble the shape of a rose. The rosettes are circular in east African leopards but square in southern African leopards.

The leopard is the smallest species in the family of big cats, but compared to its size (leopards usually weigh between 100 and 160 pounds), its strength is extraordinary. Pound for pound, it is the strongest climber of the large cats and capable of killing prey larger than itself. A leopard can climb as high as fifty feet up a tree, while holding a dead animal in its mouth, even one larger and heavier than itself! One leopard was spotted dragging a 220-pound young giraffe into heavy brush to hide it.

Leopards are nocturnal animals, meaning they are active at night. During the day, they rest in caves, thick brush, or in trees. Leopards are solitary, preferring to live alone. They can live without drinking water, getting the moisture they need from their food.

Leopards hunt at night. They use their vision and keen hearing while hunting, not their sense of smell. Leopards stalk and pounce, but don't usually chase their prey long distances. Leopards can also hunt from trees, where their spotted coats allow them to blend in with the leaves until they spring with a deadly pounce. They grab their prey or swat it, using their retractable claws. Prey is killed with a bite to the throat.

Leopards growl and spit with a screaming roar of fury when they're angry, and they purr when they're content. They announce their presence to other leopards with a rasping or sawing cough. They also leave claw marks on trees to warn other leopards to stay away from somewhere.

When it's time for a rest, leopards like to climb trees and sprawl out on the branches.

Did You Know?

Leopards can hear five times more sound than humans. They can even hear the ultrasonic squeaks made by mice.

THE CIRCLE OF LIFE

The Circle of Life

Female leopards give birth in a cave, a crevice among boulders, a hollow tree, or a thicket, in order to make a den. The average litter size is two or three cubs, although it is possible for a leopard to give birth to six cubs at once, as well. The newborn cubs are grayish, with barely visible spots. Leopards are born blind and helpless, weighing less than two pounds. Their eyes do not open until four to nine days after birth.

The female leopard hides her cubs and moves them from one safe location to the next for the first eight weeks, until they are old enough to begin playing and learning to hunt. She nurses them for three months or longer, but begins giving them meat when they are six or seven weeks old.

At around three months of age, the cubs begin to follow the mother on hunts. At one year of age, they can probably fend for themselves, but remain with the mother anyway until they are about 18–24 months old. Males take no part in the rearing of cubs.

Leopard cubs like to play "stalk, pounce, and chase." Have you ever seen a house cat creep slowly after a bird or mouse? That's stalking. A quick leap and a grab with the claws is a pounce, and the chase comes if the prey gets away. Leopard cubs play by practicing these behaviors on their brothers, sisters, and even on their mother. It's a good way for them to learn how to survive when they get older.

Leopards can live up to 15 years in the wild and up to 23 years in captivity, although 40-50 percent of cubs do not reach adulthood.

What's for Supper?

Leopards are carnivores, meaning they eat other animals. They are sly and resourceful hunters and have a very diversified diet. They will eat any meat item they can find, including monkeys, baboons, rodents, reptiles, amphibians, fish, antelope, cheetah cubs, and porcupines. They also like to eat birds and insects.

In Africa, mid-sized antelopes provide a majority of their prey, especially impala and Thomson's gazelles.

Both lions and hyenas will take away a leopard's kill if they can. To prevent this, leopards store their larger kills in trees where they can feed on them in relative safety.

Baby Fact:

The spots on the leopard's hide start out as merged dots, separating and becoming distinct as the leopard grows older.

Unique Traits

The most secretive and elusive of the large carnivores, the leopard is also the shrewdest. Admired for its distinctive beautiful fur, the leopard is cunning, alert, fearless, and bold.

The leopard has extraordinary attack weapons. It is a very agile climber and is the most accomplished stalker of all the big cats. Unlike other cats, leopards are strong swimmers and are one of the few cats that like water. They are great athletes, able to run in bursts up to 36 mph, leap twenty feet forward in a single bound, and jump ten feet into the air!

The leopard also has excellent hearing and eyesight, and is able to detect the slightest movement from half a mile away. When it moves, its heavily cushioned feet make it seem as though it is gliding rather than walking.

The leopard is believed to be more intelligent than other big cats and often employs all sorts of clever tricks to get its prey.

A British hunter watched a leopard prepare for its stalk of a buffalo calf by first rolling in buffalo dung in order to disguise its body scent. This way, it could get closer to the calf without frightening it.

Another hunter related how the leopard took a camel by arousing its curiosity. The leopard rolled on the ground, twisting and turning until it got closer to the camel. When the camel lowered its head to examine the strange animal, the leopard seized the camel and killed it immediately.

Master of Disguise

The leopard is a master of camouflage. It is extremely stealthy and is well known for its ability to go undetected. Its coat of yellow with black spots is the perfect colors for hiding in the shadows of a forest, and it blends in so much with the leopard's surroundings that even the trained eye of an experienced hunter cannot easily detect the big cat's presence. This ability to be camouflaged helps the leopard hide from its prey.

Leaping Leopards!
Snow leopards are the world's greatest leapers. They can jump as far as 50 feet!

Wacky Fact:
Leopards will swim to capture and eat crabs.

Torah Talk

The leopard isn't the largest of the big cats. It's not feared as a king of beasts. It's not the fastest, either. Rather, the leopard is persistent. It is also intelligent and strong, and is an extremely resilient and adaptable hunter, which is reflected by the huge area through which it is dispersed.

Our Sages of blessed memory describe the leopard as *"az."* In *Pirkei Avos* (5:23), Yehuda ben Taima advises us to be *az k'nameir,* bold as a leopard, in order to carry out the will of our Father in Heaven.

Elsewhere, *"az,"* or rather its derivative, *"oz,"* is translated as "might." In *Tehillim* (29:11), we are told, *Hashem oz l'amo yitein, Hashem yivarech es amo ba'shalom* — "Hashem will give might to His people; Hashem will bless His people with peace."

What is *az-oz*? What leopard-like quality are we supposed to have, according to Yehuda ben Teima? Let's look back to the leopard for help with answering this.

The most famous leopard in Tanach is in the messianic vision of Yeshayah Hanavi (*Yeshayah* 11:6), where the predator leopard lies down with a young goat.

The leopard is also seen as part of the oppression of the Jewish people. Rashi says that the leopard, the third creature in Daniel's vision (*Daniel* 7:6), represents the evil kingdom of Antiochus, whose decrees were "spotted."

R' Gershon Winkler, author of *Soul of the Matter*, says that the leopard symbolizes stalking and patience, which then leads to sudden transformation. Its power is in its sudden appearance, its element of surprise. This is symbolic of redemption.

Putting all of this together, we have solitude, persistence, dispersion, resilience, patience, strength, intelligence, boldness, war, oppression, surprise, and redemption. We have the story of the Jewish people, who persist despite the world's attempts to be rid of us.

The key to our might, the basis for our persistence, lies in doing the will of our Father in Heaven. If we follow the ways of the Torah, then, like a leopard, sudden transformation and redemption will come upon us. Let Klal Yisrael be *"az k'nameir"* to serve Hashem. That way, we'll truly merit Hashem's blessing of peace.

Record Holder:
The world's rarest big cat is the Amur or Manchurian leopard of which fewer than 35 are believed to exist. It is currently classed as Critically Endangered by the IUCN.

INTERESTING FACTS & STATS

🐾 Leopards can drag up to three times their own body weight into a tree and place it on branches almost twenty feet high.

🐾 Leopards are 1.5–2.6 feet tall at the shoulder. They are three to six feet long, and weigh between 82 and 200 pounds.

🐾 Leopards have thirty-two teeth, four of which are long, pointed canine teeth.

🐾 The leopard is the fifth largest feline in the world behind the tiger, lion, jaguar, and mountain lion.

🐾 Each leopard's spots are unique, similar to human fingerprints.

🐾 Leopards have several extra long hairs in the eyebrows to help protect the eyes and assist them in moving through vegetation in darkness.

🐾 The leopard's whiskers, as with all cats, have specialized sensory hairs that register minute changes. This enables the leopard to avoid objects in the dark.

🐾 Female leopards are twenty to forty percent smaller than males.

🐾 The white spots on the tip of their tails and back of their ears help leopards locate and communicate with each other in tall grass.

🐾 King Nimrod was a great-grandson of Noach. He subdued a leopard, which then accompanied him on his hunts. According to some, the word "Nimrod" is Babylonian for "leopard-tamer."

Leopard Trivia

1. How long is the tail of a leopard?
a. 8–13 inches **b.** 15–20 inches **c.** 24–43 inches **d.** 48–60 inches

2. With which animal is the leopard sometimes confused?
a. jaguar **b.** lion **c.** hyena **d.** tiger

3. Which one of these statements is false?
a. Leopards are good swimmers. **b.** Leopards use their sense of smell to hunt. **c.** Leopards are nocturnal.

4. The leopard flips its tail over its back and reveals its white underside when it is…
a. hungry **b.** thirsty **c.** ready to attack **d.** giving a sign that it is not seeking prey

5. Where do leopards usually hide their food?
a. nowhere — they eat it immediately **b.** in their den **c.** up a tree **d.** underground

ANSWERS: 1. c 2. a 3. b 4. d 5. c

Animal Crackers:
Q. How does a leopard change its spots?
A. When it gets tired of one spot it just moves to another.

LION　　אריה

The Lion

The lion is one of the largest, strongest, and most powerful felines in the world. They are the largest cats on the African continent and are unique among felines in a number of ways; however, the biggest difference between lions and other big cats is that only lions, being incredibly sociable animals, live together in family groups, which are known as prides. A pride is made up of five to fifteen related females and their cubs, along with (generally) a single male (although small groups of two or three lions are not uncommon).

Despite their enormous size, male lions hardly do any of the hunting, as they are often slower and more easily seen than their female counterparts. The lionesses in the pride do the hunting, usually together. They are quite successful on their trips; lionesses are even able to catch and kill animals that are both faster and much bigger than them.

Lions have a short coat of tawny or golden fur, with a long tail that has a tuft of longer fur at the end. The markings on their coats are much fainter than the bold stripes and spots displayed on other felines, which helps the lions go unseen when stalking prey in the tall grasses. Male lions are taller and heavier than the females and have a mane of long hair around their faces (in fact, it is the only case in the feline world where males and females actually look different).

Lions have strong and powerful jaws that contain thirty teeth in total. This includes their four fang-like canines and four carnassial teeth that are especially designed for slicing through flesh. Lions also have large paws with soft pads underneath, and a sharp, retractable claw on the end of each toe; these aid these large carnivores in running, climbing, and catching their prey. The structure of a lion's feet and legs enable it to jump distances of over thirty feet!

Lions are incredibly adaptable animals that can and will inhabit very dry climates, as they get most of the moisture they need from their food. They prefer areas of open woodland, scrub, and long grasslands, where there is not only plenty of cover but also a wide variety of prey.

The lion is the most dominant predator within its environment. This means that other animals pose little or no threat to them, with the exception of hyena packs, which can cause fatal damages to lions, particularly when a lion is on its own and food is scarce.

Did You Know?

The roar of an adult lion can be heard over five miles away.

THE CIRCLE OF LIFE

The Circle of Life

Female lions give birth to between one and six cubs, all of which are born blind and are incredibly vulnerable in their new surroundings. The lion cub only weighs between two and four pounds at birth. The fur of the lion cubs is covered in darker spots that help to camouflage them in their den in order to protect them while the adults go out to hunt. Less than half of cubs make it to be a year old, and four out of five will die by the time they are two, generally either from animal attacks or starvation.

Lion cubs nurse from their mother until they are about six months old, and although they won't begin actively hunting until they are about one year old, they start to eat meat after twelve weeks or so. The males generally begin to develop manes when they are three years old.

Young lion cubs spend a great deal of time playing together, which actually helps them to develop their hunting techniques. This method of role-playing among the cubs also helps females to determine what the cubs would be best suited for: chasing and cornering prey, or catching and killing it. (The lionesses are the ones that allocate specific hunting tasks to each cub.)

The mother lion will raise her male cubs until they reach two years of age; after that, the males are forced out of the group and are on their own. These males generally form small bachelor groups and roam. The female cubs, however, remain with the pride, and may even live together with their mother for their entire lives.

Lions can live up to fifteen years in the wild.

What's for Supper?

The lion is a carnivorous animal that survives only by eating other animals. Unlike other felines, lions are not solitary hunters; instead, the lionesses work together in order to catch their prey, with each female having a different strategic role. This strategy allows them to kill animals that are both faster and much larger than they are, including buffalos, wildebeests, and even giraffes. Lions primarily hunt gazelles, zebras, and warthogs, along with a number of antelope species, by following the herds across the open grasslands.

Baby Fact:
Many females in a pride give birth at about the same time as each other. Remarkably, the cub will nurse from any lioness in the pride, not just from its own mother.

UNIQUE TRAITS

Unique Traits

The mane of the adult male lion may be one of the most distinctive and majestic physical traits among all the creatures in the animal kingdom. The male lion is the only member of the cat family with a mane. This mane is long, thick, and furry, and covers the lion's head and neck. In some lions, the mane may even extend to the abdomen. The color of the lion's mane varies and may be blonde, brown, or black. Although the mane does give the male lion a kingly appearance, it is not only meant for show; the mane serves many other functions, as well.

The length and color of the mane is believed to be an indicator of many factors, including the lion's age, health, genetics, and hormones. Generally, the darker and fuller the mane, the healthier and more mature the lion is.

The main responsibility of the male lion is to protect the pride. An impressive mane can make the lion appear bigger and more powerful than it really is. This intimidating appearance helps to deter attacks from its enemy, the spotted hyena. An impressive-looking mane also helps to avoid battles with other male lions looking to take control of the lion's pack, as lions will think twice before starting up with another lion that has a greater mane than they do.

When a lion does engage in battle with another lion, its thick mane acts as a defensive shield, as it protects the lion's neck and throat against the striking claws and biting teeth of its enemy.

Research has suggested that environmental factors, such as the temperature, influence the color and size of a lion's mane. It should be noted that some male lions do not have manes at all, especially those living in hot, dry, and arid climates. Cooler temperatures seem to result in a heavier mane. Zoo lions have larger manes than is typical of lions that live in the wild. Better nutrition, less wear and tear, and regulated climatic effects could all contribute to this difference.

Wacky Fact: When lions walk, their heels don't touch the ground.

Torah Talk

Be as strong as the lion to do the will of your Father in Heaven (Avos 5:20).

Initially, this teaching may seem rather strange. How can there be any comparison between the lion, which can be cruel and ferocious, and the will of Hashem, Who is kindly, merciful, and gracious? However, the Hebrew word for strong, *gibor*, does not refer to the exceptional physical strength of the lion, but to the ability with which Hashem has endowed it to curb its passions. This is in the spirit of the teachings of Ben Zoma: *Who is mighty (gibor)? He who subdues his passions (Avos 4:1).*

The lion will devour and plunder when it is hungry and will provide for the needs of the members of its family, but it is not naturally cruel. In contrast to the bear, the lion will not recklessly hunt for prey. The lion has been known to aid weaker animals and even procure food for them. Indeed, the lion has the strength to subdue its passions. This may explain why the righteous and upright of Klal Yisrael are *stronger than lions to do the will of their Master and the desire of their Rock (Av Harachamim, Shabbos morning prayer).* This evaluation of the "might" of the lion is borne out by our Sages in the Gemara, where we learn that sometimes the lion will stay among the flocks and not injure them (*Chullin* 53a).

A story is recorded in the Gemara about the lion. Rabbi Shimon ben Chalafta was walking on the road when lions met him and roared at him. Rabbi Shimon responded by quoting the verse, *The young lions roar after their prey and seek their food from Hashem (Tehillim* 104:21). Immediately, two lumps of flesh descended from Heaven, whereupon the lions ate one and left the other (*Sanhedrin* 59b). We see again how lions can curb their appetites and subdue their passions.

It is in this spirit that the lion is called the king of the beasts (*Chagigah* 13b). It has earned this title because it can govern its passions. In this respect, some authorities identify the thick, shaggy mane around the head of the lion as its crown.

The lion is the only member of the cat family with a tasseled tail. It uses it to signal commands, such as: "this way" or "come here."

Record Holder:
The heaviest lion on record was a male lion named Simba. It weighed a whopping 826 pounds!

INTERESTING FACTS & STATS

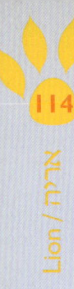

🐾 The average male lion weighs around 400 pounds, and the average female lion weighs around 290 pounds.

🐾 Lions can reach speeds of up to 50 mph, but only in short bursts, due to a lack of stamina.

🐾 The lion has a good sense of hearing and is able to hear its prey from one mile away.

🐾 Lions can smell nearby prey and estimate how long it was in the area.

🐾 Tigers are so similar to lions that without their coats, only experts are able to tell them apart, as their bodies are almost identical.

🐾 When a new male lion joins a pride, it will usually kill the cubs.

🐾 Lions are able to go for four days without drinking water.

🐾 One way zoologists identify individual lions is by recording the spots on their muzzles.

🐾 A lion has retractable claws.

🐾 Lionesses usually hunt in the dark. They often hunt in groups of two or three, using teamwork to ambush and kill their prey.

🐾 A lion can eat around forty pounds of meat in one sitting.

🐾 A single lion has only about an 18 percent chance of catching prey, while a group of lions has around a 30 percent chance.

Lion Trivia

1. **For how many hours during the day does a lion usually rest?**
a. four b. eight c. sixteen d. twenty

2. **The only animal that poses a real threat to the lion is the…**
a. tiger b. grizzly bear c. hyena d. crocodile

3. **At what age does a lion start to roar?**
a. one week old b. one month old c. one year old d. two years old

4. **Which statement is false?**
a. Lions have excellent eyesight. b. All male lions have a mane.
c. All lions have a tuft at the end of their tails. d. Lions are good swimmers.

5. **Besides lions, what other cats live in groups?**
a. tigers b. tigers and leopards c. jaguars d. none — lions are the only ones

ANSWERS: 1. d 2. c 3. d 4. b 5. d

Animal Crackers:
Q. What does the mother lion say to her cubs before they go out hunting for food?
A. "Let us prey."

MONKEY קוף

Banana Split

Monkeys do not eat the skin of a banana. Before digging in, they first peel the banana — from the opposite end!

Fast Fact:
Howler monkeys have special structures in their throats that are oval-shaped and hollow. These structures act as echo chambers and help make the monkeys' calls extremely loud.

The Monkey

Monkeys are found naturally in the jungles and forests of the Southern Hemisphere. There are around 260 known species of monkeys worldwide. They come in all shapes, colors, and sizes (the size of a monkey can range from just a few inches long to more than three feet tall), and are divided into two groups: the Old World monkeys, which are found in Asia and Africa; and the New World monkeys, which are found in South America.

The Old World monkeys are larger than the New World monkeys. Although both the New World monkeys and Old World monkeys have forward-facing eyes and share some features, such as the types of noses and cheeks, the faces of the two kinds of monkeys look very different.

In this section we will explore two types of Old World monkeys and two types of New World monkeys.

The Howler Monkey

Howler monkeys are New World monkeys, and are dispersed throughout the tropical jungles of Central and South America. They are aptly named for their piercing, howling cries. The howler monkey is thought to have the loudest call of all the primates in the world, with some howler monkeys being able to project their howling voices for up to a few miles!

The howler monkey is the largest of all the New World monkeys, with some species growing a little more than three feet tall. Yet despite their large size, howler monkeys weigh less than twenty-two pounds, which allows them to move with more agility through the high trees.

Howlers, as well as all the other New World species, have wide, side-opening nostrils and prehensile tails. They use their tails as an extra arm with which to grip or hang from branches. A gripping tail is especially helpful to howler monkeys, since they rarely descend to the ground. They prefer to stay high in the trees, munching on the fruit and leaves that make up most of their diet.

Howler monkeys move in groups of about eighteen monkeys per troop. They spend most of their time sleeping and grooming each other. The howler monkey is said to be one of the least active monkeys, as it spends eighty percent of its time resting! Howler monkeys generally live to around twenty years of age.

Baby Fact:
Several female howler monkeys often help a mother take care of her baby.

What Was That?
Because of their speed and coloring, Patas monkeys have often been mistaken for cheetahs while running.

Did You Know?
Troops of Patas monkeys never spend two nights in the same place.

The Patas Monkey

The Patas monkey, also known as the military monkey, the Hussar monkey, and the red guenon, is a medium- to large-sized species of the Old World monkeys which lives in the open grasslands of Central Africa. It has long limbs, large hands, and big feet, and is able to run at great speeds. It is generally found in areas with very little cover, and so, if threatened, all it can do is simply run away. The Patas monkeys' long back legs are so powerful that they are able to reach speeds of up to 35 mph, making these monkeys the fastest primates in the world.

The Patas monkey has a long and slimly built body covered in shaggy fur, which is white on the underside and red on the back. Its long limbs are also white, while its face is dark, with a white moustache and beard, and a red cap with a heavy brow ridge that protects its eyes. Patas monkeys also have a distinctive black line that runs from their eyes up to their ears, which are near the top of their heads. Due to their smart red coats and soldier-like white moustaches, Patas monkeys are commonly known as "military monkeys." Although male and female Patas monkeys look remarkably similar, the males tend to be much larger in size and also have a slight bump that protrudes from their heads.

Unlike numerous other primate communities, Patas monkey troops are led by the females. There are between ten and forty members in a troop, with only one adult, dominant male. (All other males must leave the troop at four years of age.) The role of the male Patas monkey is to protect the troop from danger. Since Patas monkeys live in the open country, the troop is easily viewed by predators like lions and leopards. The male Patas monkey lingers on the outskirts of the troop and watches out for approaching danger. He will stand on his feet to look out from the tall grass, or climb into a tree and watch from there. There are times that he will act as a decoy to distract predators and run in the opposite direction of the troop, allowing the females and the young to run off and hide.

The Patas monkey is an omnivorous animal that consumes a wide range of both plant matter and small animals in order to survive. More than eighty-five percent of its food is collected on ground level. It feeds primarily on fruits, leaves, flowers, and tree gum. It is also known to eat insects, lizards, and birds' eggs.

Wacky Fact:
Patas monkeys can hold as much food in their cheeks as they can in their stomachs!

What's in a Name?
The word mandrill means "man-ape."

Fast Fact:
When a male mandrill gets angry or excited, the colors on his face become even brighter. This serves as a warning signal to others.

The Mandrill

The mandrill is a powerful Old World monkey and is considered to be one of the largest monkey species in the world. It is rather shy and seclusive and can only be found in a small pocket of tropical jungle in western Central Africa. The mandrill is very colorful, perhaps more so than any other mammal. It is easily recognized by its face's blue and red skin. The mandrill has a long snout; sharp teeth; a thick body; and arms that are longer than its legs. Its tail is just a short stump, while its cheeks have built-in pouches that are used to store snacks for later consumption. Though mandrills spend most of their time on the ground, they do climb trees and even sleep in them.

The adult male mandrill is nearly three feet tall. It has an orange-yellow beard and a large head that is topped with shaggy, brown hair. Its long, bluish-white snout has a striking, bright red streak right down the middle. A male mandrill's color is very significant. The more colorful it is, the more successful it is within a mandrill troop. The males with the most colorful snouts are the most dominant. They lead the troop and protect it from enemies.

The male mandrill also has incredibly long teeth, which it bares as a warning to predators. Besides for using its teeth to protect itself against enemies, it will show off its teeth as a friendly gesture among fellow mandrills.

Mandrills inhabit areas of the forest in large troops. Troops consist of one alpha male and fifteen or more females and their young. Mandrills also gather in multi-male/multi-female groups that can consist of over two hundred members.

Mandrills are omnivorous animals and will eat almost anything. They mainly feed on fruits, berries, seeds, nuts, roots, leaves, insects, and even small mammals and reptiles. Most of their diet is found at ground level or just above.

Due to their large size, mandrills have few predators in their natural environment. The leopard is the main predator of the mandrill, although young mandrills need to watch out for large snakes and birds of prey, too.

A mandrill can live for an average of twenty-eight years in the wild.

Did You Know?

The largest mandrill group, found in Lopé National Park, Gabon, had 1,300 mandrills. It is the largest gathering of monkeys ever recorded.

Finders Keepers...
Squirrel monkeys follow other troops of monkeys at a distance, to take advantage of the food left behind.

Baby Fact:
Shortly after giving birth, the female squirrel monkey will chase away the male, who plays no part in raising the baby.

The Squirrel Monkey

The squirrel monkey is a small species of the New World monkeys that lives in the forests and jungles of Central and South America. Measuring as little as ten inches from the top of its head to the base of its tail, this tiny primate is more than double that size when its long tail is included. Unlike a number of other small monkey species, the tail of the squirrel monkey is not prehensile, which means that it cannot be used to grip onto branches. Instead, this long tail is used to help the squirrel monkey balance when climbing on high branches.

The squirrel monkey has a very distinctively colored, short fur, which is mostly olive or gray in color, with the exception of its bright yellow legs and white face. It also has a tuft of longer and darker hair on its forehead, and a black or dark brown muzzle. It spends a great deal of time high in the trees, and has dexterous fingers that are not only great for gripping branches, but also prove to be very useful when opening fruits and holding onto prey. Squirrel monkeys are excellent at climbing and leaping between branches as they travel through the forest.

The squirrel monkey is thought to be one of the most intelligent species of primates and is known to have the largest brain-to-body mass ratio of all the monkey species in the world. Squirrel monkeys have incredibly good eyesight and color vision, which means that they are able to spot fruits among the dense vegetation with ease.

Squirrel monkeys move about noisily in the trees in large troops that are usually forty or fifty animals strong, but can contain up to five hundred monkeys. Troops usually contain a number of sub-groups, including adult males, expectant females, females with their young, and groups of young squirrel monkeys. The whole troop sleeps together at night, and then breaks up into its various sub-groups to feed during the day.

Squirrel monkeys use a number of vocal calls to communicate with each other, including special warning sounds that indicate the presence of a predator. As these monkeys are from the smallest species of New World monkeys, they are preyed upon by a variety of forest animals.

Fast Fact:
While foraging for food in the thick foliage of the jungle, squirrel monkeys make a "chuck-chuck" sound to indicate their location to the other members of their troop.

Unique Traits

The monkey is considered to be one of the most intelligent animals on Earth. Many monkey species display complex social behaviors in their daily lives, from the way they care for their young to their usage of basic tools in order to obtain food for themselves.

Monkeys use facial expressions, body movements, and noises called vocalizations, to communicate with one another. Staring, for instance, is a threat in monkey communities. Grinning or showing the teeth signals aggression or anger, while grooming is a sign of friendship and peace. Vocalizations can be quite complex, and are often used to warn other monkeys of predators nearby. There is research that suggests that some monkey species have developed methods of putting together phrases and are actually able to "talk" to one another!

There are organizations that train certain species of monkeys, such as the white-faced capuchin, to be "monkey helpers," in order to assist quadriplegics and other people with mobility impairments. After being raised in a human home as an infant, the monkey will then undergo extensive training before being placed with a quadriplegic. Once the monkey has been trained, it helps out around the house by doing tasks such as microwaving food and opening drink bottles!

The traits of the white-faced capuchin are quite fascinating. Because of its particular eating habits, creative use of tools, and complex social structures, it is considered to be the most intelligent of the New World monkeys. One way the white-faced capuchin exhibits its intelligence is when searching for food; it looks under logs and foliage for insects, and can be seen squeezing and smelling fruit to check if it is ripe. It also uses stones and twigs to pry open shellfish, in order to get to the meat inside.

Another species of monkeys, the rhesus monkey, is considered to be the most intelligent of the Old World monkeys. The rhesus monkey possesses a remarkable memory, as well as the ability to learn and to make decisions. Because of its high intellect and its similarity to humans anatomically and physiologically, the rhesus monkey has been used in biological and medical research. It was instrumental in the discovery of the Rh (rhesus) factor in blood and the development of the rabies, smallpox, and polio vaccines.

Flying Monkeys?
The first monkey astronaut was Albert, a rhesus monkey, who, on June 11, 1948, rode on a V2 rocket for six minutes.

Wacky Fact:
A monkey was once tried and convicted for smoking a cigarette in South Bend, Indiana.

Torah Talk

The Gemara in *Brachos* (58b) teaches that one must recite a *brachah* upon seeing certain unusual creatures; specifically singled out as "unusual creatures" are monkeys and elephants. These rulings are codified by the Rambam (*Hilchos Brachos* 10:12–13) and by the *Shulchan Aruch* (*Orach Chaim* 225:8, 10), which states that the *brachah* to be recited upon seeing unusual creatures is *Meshaneh Habriyos* — "Who has altered the creations." In practical halachah, Rav Ovadiah Yosef (*Sefer Yalkut Yosef*, Volume 3, 225:21) rules that one should recite the *brachah* of "*Meshaneh Habriyos*" upon seeing a monkey or an elephant.

The obvious question is: Why are monkeys and elephants the only animals to be singled out as being "unusual creations"? Certainly, there are countless other animals that have their own unique and often strange appearances!

The reason only these two animals are singled out for this *brachah* is as follows:

The midrashim (*Bereishis Rabbah* 38) tell us that the sin of *Migdal Bavel* was a conspiracy made among the people to rebel against their Creator. Hashem punished the people by scattering them across the planet in order to break their unity, which had turned to evil.

Among those who sinned, however, there were different groups, each of which was punished differently. Those who had wanted to sit quietly atop the tower they had built were dispersed all over the world. Those who had intended to use the tower for idolatry lost their common language and unity. Those who had wished to use the tower to do battle against Hashem were changed into monkeys and demons (*Sanhedrin* 109a).

As for the elephants, the *Meleches Shlomo Kil'ayim* (8:6) teaches that in the generation of the *Mabul*, some of the people of the generation turned into monkeys, and some people turned into elephants.

It is for this reason that the halachah says that when one sees a monkey or an elephant, he should recite a special blessing to the One "Who has altered the creations" — since these animals came about as a result of the creations being changed!

Record Holder:

On May 28, 1959, Able, a rhesus monkey, and Miss Baker, a squirrel monkey, were fired 300 miles into space in the Jupiter missile AM-18 from Cape Canaveral in Florida. They were the first monkeys to successfully return to Earth alive after traveling in space.

INTERESTING FACTS & STATS

- Howler monkeys usually make their loud booming calls in the morning and at the end of the day.
- Howler monkeys get almost all the water they need from the food they eat.
- Despite its large size when an adult, the mandrill weighs less than two pounds at birth.
- Female mandrills form a life-time bond with their daughters; however, their sons leave them once they reach maturity.
- Patas monkeys walk on their fingers instead of their palms.
- Patas monkeys will lean back and put up their feet when relaxing!
- Newborn squirrel monkeys cling to their mothers' fur and travel around with them as soon as they are born.
- Squirrel monkeys have nails instead of claws.
- The pygmy marmoset is the world's smallest monkey. It measures four and a half to six inches in length and weighs three to five ounces.
- The rhesus monkeys preceded humans into space.
- Capuchin monkeys use different noises to identify different types of predators. They will also bang rocks together to warn each other of approaching danger.
- Apes and spider monkeys swing arm-to-arm in trees; however, most monkeys do not travel by swinging from tree to tree. Instead, they travel through the treetops by running across the branches.

Monkey Trivia

1. Which characteristic do Old World monkeys and New World monkeys share?
   a. They have prehensile tails. **b.** They have pouches in their cheeks to store food.
   c. They have forward-facing eyes. **d.** They live in the jungles of Africa.

2. Which monkey is the least active?
   a. the mandrill **b.** the howler monkey **c.** the Patas monkey **d.** the rhesus monkey

3. Which monkey is the largest?
   a. the mandrill **b.** the squirrel monkey **c.** the Patas monkey **d.** the white-faced capuchin

4. The primary function of the squirrel monkey's long tail is…
   a. for balance **b.** to swing from tree branches with **c.** to grab food with **d.** to use as a weapon

5. Which monkey is called "the dancing monkey" because it jumps when it's excited?
   a. the mandrill **b.** the howler monkey **c.** the squirrel monkey **d.** the Patas monkey

ANSWERS: 1. c 2. b 3. a 4. a 5. d

Animal Crackers:
Q. Why did the monkey go to the doctor?
A. Because his banana wasn't peeling very well!

PORCUPINE דרבן

The Porcupine

The porcupine is the prickliest of all rodents. There are about two dozen porcupine species, and all boast a coat of needle-like quills to give predators a sharp reminder that this animal is no easy meal. Some quills, like those of Africa's crested porcupine, are nearly a foot long!

Porcupines are generally black or brown. The quills are lighter in color, often with black and yellow or black and white alternating rings. They have soft hair, but on their back, sides, and tail, the hair is usually mixed with sharp quills. These quills typically lie flat until a porcupine is threatened; when that happens, the quills stand up straight as a means to keep the threat away.

Biologists separate porcupines into two main categories: *Old World porcupines* and *New World porcupines*. Old World porcupines live in Africa, Europe, and Asia. These animals usually live on the ground and can inhabit deserts, grasslands, and forests. New World porcupines are found in North, Central, and South America.

The most common New World species is the North American porcupine, also known as the common porcupine. These porcupines are good climbers and spend most of their lives in trees. Some even have prehensile (gripping) tails that can grasp branches like monkeys.

The North American porcupine is the only species that lives in the U.S. and Canada, and is the largest of all porcupines. An adult is about 37 inches long, including the tail, and weighs approximately 25 pounds. Its quills can reach 8 inches long.

The North American porcupine is a solitary animal, although it may den with other porcupines in the winter. It makes its den in caves, decaying logs, and hollow trees.

The porcupine is the only native North American mammal with antibiotics in its skin. These antibiotics prevent infection when the porcupine falls out of a tree and is stuck with its own quills upon hitting the ground. It is fairly common for porcupines to fall out of trees, because they are highly tempted by the tender buds and twigs at the ends of the trees' branches, and will try reaching for them — sometimes unsuccessfully.

The porcupine and the skunk are the only North American mammals that have black and white colors, because they are the only mammals that benefit from letting other animals know who and where they are in the dark of the night.

Did You Know?

Contrary to popular myth, porcupines cannot shoot their quills.

THE CIRCLE OF LIFE

The Circle of Life

After a seven-month pregnancy, the female porcupine gives birth to a single offspring. The baby porcupine, called a *porcupette*, is almost always born head first in order to protect its mother from the quills. The porcupette's quills are soft and bendable at first, but within a few minutes of being exposed to the air, they start to harden.

Porcupettes weigh only around one pound at birth, but develop very quickly and almost immediately have full use of their eyes. Within two weeks, the baby will begin eating vegetation. It will stay with its mother for just a couple of months before it is ready to live on its own.

A fully-grown adult porcupine is covered with over 30,000 sharp, detachable spines.

Having quills, however, does not necessarily mean that the porcupine's life is trouble-free. Large cats, especially lions, and human hunters threaten Old World porcupines. New World porcupines' predators include wolverines, pythons, eagles, and great horned owls. One porcupine predator, the fisher, is able to flip the North American porcupine onto its back, exposing its unprotected belly.

Porcupines are relatively long-lived animals that can live up to twenty years in the wild.

What's for Supper?

The porcupine is an herbivore, which means it only eats plant matter. This includes a variety of plants, shrubs, leaves, twigs, and herbs, with a preference for green plants such as skunk cabbages and clover. In the winter, when food is scarce, the porcupine will also eat the tender layer of tissue beneath the bark of trees. A porcupine can cause a whole tree to be knocked down if it removes too much bark.

Porcupines may gnaw on bones to sharpen their teeth and to obtain salt at the same time. As they are vegetarian animals, they do not receive enough sodium through their regular diet, and are always on the lookout for additional sources of salt. In search of salt, porcupines have also been found to eat tool handles, doors, tables, footwear, clothing, and other items that have been coated in salty sweat.

The North American porcupine often climbs trees to find food such as leaves and berries. The African porcupine on the other hand, is not a climber, and therefore searches for food on the ground.

Baby Fact:
Porcupine babies are not born with sharp or barbed quills; thank goodness for mom's sake!

Unique Traits

Even though it is true that a porcupine has a calm disposition and generally dislikes battle, once it does decide to advance, nothing in the world will stop it. It weighs only fifteen to twenty-five pounds and possesses a tail that's only six inches long, but it is armed with a terrifying arsenal of quills that sprout from its head, back, and powerfully muscular tail. When it walks, it rattles like a quiver of arrows — which is exactly what it is. The quills are hollow, tubular, and attached so lightly to its skin that the slightest touch will dislodge them. And just hope that they never come your way. The quills are as sharp as needles, and covered with a multitude of barbs. As soon as the quills enter the flesh of a victim, they begin to swell up, with the barbs sticking out more and more.

If Mr. Porcupine falls into the water, his air-filled quills keep him afloat, buoyant as a cork. If he tumbles out of a tree (his favorite habitation), his cushion of hollow quills breaks his fall and gives him a comfortable landing. Should an enemy be foolish enough to threaten him, his behavior is always the same. He brings his feet close together, hugging the ground to guard his unquilled underside. Next, he raises his quills until, like a fantastic pincushion, he looks twice his size, and vigorously flips his tail from side to side.

If the attacker is wise, he goes away. But if he continues trying to close in on the porcupine — slap! The muscular tail lashes sideways and drives as many as twenty jagged stilettos deep into the attacker's flesh. One slap is generally sufficient to drive off even a bear, but should the enemy stand his ground, the prickly porcupine advances. In order to protect its nose, the only sensitive part of its body that is exposed to the attacker, it goes into reverse gear and advances backward, flailing its tail furiously. Ten quills will drive off a fox and twenty will send a wildcat away screaming in pain; in fact, both mountain lions and bears have been killed by porcupine quills.

But doesn't the porcupine ever run out of ammunition? Have no fear; its lost quills are replaced by new ones within only a few months. In the meantime, the porcupine is unlikely to run short, for it has no fewer than 30,000 quills covering the length of its little body.

Who instructed the porcupine to manufacture quills that expand inside their hapless victim? Who is the Designer behind this advanced weaponry? It is the greatest Intelligence, *Hashem Yisbarach*, Who created this world with the wisdom that is everywhere, if one only makes the effort to look for it.

Ouch! Native Americans used the quills of the porcupines for hair brushes!

Wacky Fact: All porcupines can float in water — their quills act as a life preserver!

Torah Talk

The *pasuk* in *Yeshayah* mentions a wild creature named a *kippod* which lived in desolate, ruined places (*Yeshayah* 14:23; 34:11). Some commentators interpret this creature to be the porcupine.

Yeshayahu predicted the destruction of the first Beis Hamikdash in vivid detail, and is best known for his prophecies of consolation and redemption, which are read as the *haftaros* on the Shabbosos following Tishah B'Av. The *haftorah* that is read on the morning of Tishah B'Av, however, addresses why Hashem punished the Jews:

A slaughtering arrow is their tongue, deceit they speak with their mouths: they speak peaceably to their friend, and inside [their hearts] they plant their ambush. On these [people] shall I not visit them [with punishments], so says G-d; if with a nation as this, shall My soul not take revenge? (Yirmiyah 9:7, 8).

We learn from here that baseless hatred, *sinas chinam*, was the cause for the Beis Hamikdash's and the Jews' destruction.

By observing the porcupine, perhaps we can learn some valuable lessons regarding our personal relationships with our peers. Below is a story told by Rabbi Simcha Weinberg that helps illustrate this point:

A Message about *Sinas Chinam* — Baseless Hatred

It was the coldest winter ever. Many animals died because of the cold. The porcupines, realizing the situation, decided to group together to keep themselves warm.

Even like this, though, the porcupines weren't happy, because now their quills were wounding each other. So, after a while, they decided to separate.

Soon enough, the porcupines began to die, alone and frozen. They realized now that they had to make a choice: either to accept the quills of their companions, or to die.

Wisely, they decided to go back to being together again. They learned to live with the little wounds caused by the close proximity of their companions, in order to receive the warmth that came from them. In this way, they were able to survive.

Moral of the story: The best relationship is not the one that brings together perfect people, but when each individual learns to live with the imperfections of others and can admire the other person's good qualities.

Record Holder:
On August 17, 2011, an Indian crested porcupine by the name of Ferdinand became the oldest living porcupine by marking its 30th birthday at the Prague Zoo.

INTERESTING FACTS & STATS

🐾 There is an archeological site in Israel called Horvat Darban. It acquired its name when a porcupine (*darban* in Hebrew) was digging itself a burrow in the southern slope of the ruins, and exposed a potsherd bearing an imprint that stated "for the king." The potsherd dated from the time of King Chizkiyahu (705–701 BCE).

🐾 Quills are just modified hairs made out of keratin, the same substance found in our own hair and fingernails.

🐾 Porcupines may look awkward on land but they are good swimmers.

🐾 Porcupines are nocturnal, which means they're active at night and sleep a lot during the day.

🐾 New World porcupines are also known as tree porcupines.

🐾 Porcupines grow new quills to replace the ones they lose.

🐾 Female porcupines have between one and four young in their lifetime, depending on the species.

🐾 Africa's largest rodents are porcupines.

🐾 To chomp away at their food, porcupines have sharp, chisel-like front teeth.

🐾 Newborn porcupines weigh about three percent of their mother's weight at birth.

🐾 Porcupines tend to grunt when foraging for food.

🐾 Porcupine quills are hollow and have spines on the ends that make them hard to pull out of a porcupine's victim.

Porcupine Trivia

1. Another name for the porcupine is…
a. porcy **b.** quill pig **c.** prickly rodent **d.** needle rat

2. What do porcupines like best?
a. salt **b.** mushrooms **c.** berries **d.** ants

3. What is the porcupine's worst enemy?
a. humans **b.** the bear **c.** the lion **d.** the fisher

4. Which one of these statements about porcupines is false?
a. They have a good sense of smell. **b.** They have good eyesight. **c.** They have good hearing.

5. The porcupine is the second largest rodent in North America. Which is the largest rodent in North America?
a. hedgehog **b.** rat **c.** beaver **d.** elephant

ANSWERS: 1. b 2. a 3. d 4. b 5. c

Animal Crackers:
Q. What do you get when you cross a turtle and a porcupine?
A. A slowpoke

SNAKE נחש

Snake Eyes!
The rattlesnake stores its poisonous venom in glands below and behind its eyes.

Baby Fact:
Rattlesnakes do not lay eggs in nests. Instead, they give birth to live young.

Snakes

Snakes belong to the animal group called reptiles. Snakes have long, thin bodies; no arms or legs; no eyelids; no ears; and are covered with back-folded skin sections called scales. There are around 3,000 known species of snakes worldwide. Snakes are found on land and in water on every continent except in the polar regions, where it is too cold for them.

All snakes are carnivores. They usually kill their prey in one of two ways: by delivering a poisonous bite to it, or by wrapping themselves around their prey in order to constrict it. In this section, we will take a closer look (don't worry — we won't get too close!) at two popular, poisonous snakes and two well-known constrictor snakes.

The Rattlesnake

There are more than thirty species of rattlesnakes, ranging in sizes of one foot to eight feet in length. All rattlesnakes are poisonous and can be recognized by their distinct sound of…the rattle! The rattle, which is made up of a series of hollow, bony segments of keratin, is located at the tip of the rattlesnake's tail and is used to warn predators to keep away. The famous rattle noise is created by the contraction of special "shaker" muscles in the tail, which cause these bony segments to vibrate against one another.

Rattlesnakes are also called pit vipers. Their triangular heads contain a hollow spot between their eyes and nostrils called a pit, which is actually a heat-sensitive sensory organ that helps the rattlesnake locate its prey (such as a rabbit or a mouse) in darkness, by detecting the prey's body heat. This enables the snake to actually "see" a heat image of its prey — even in complete darkness! Rattlesnakes are also able to detect movement of their prey by sensing their vibrations on the ground.

Rattlesnakes kill their prey with a poisonous bite, rather than by constricting it. When the rattlesnake bites its prey, it injects venom from its fangs into it. The venom, which contains powerful digestive enzymes, starts digesting the prey from the inside before the snake even swallows the animal.

Female rattlesnakes only give birth once every two to three years (usually having ten to twenty babies at once). After one to two weeks, a newborn rattlesnake sheds its skin and grows the first segment of its rattle. The baby rattlesnake leaves its mother at just a few weeks of age, but returns to its mother's den when it is time to hibernate in the winter.

The average lifespan of a rattlesnake is twenty-five years.

Did You Know?
The Mojave rattlesnake has neurotoxin in its venom, making it the most dangerous of all rattlesnakes.

(Not So) Fast Fact:
Pythons can travel only about one mile per hour on the ground.

Fast Fact:
Many pythons are excellent swimmers and spend a lot of time in the water.

The Python

Pythons are among the largest snakes in the world. There are over thirty species of pythons. Most live mainly on the ground and occasionally climb trees, although some species live only in trees, and a few species are burrowers.

The python is not poisonous; however, it can attack and cause severe injuries with its long, sharp teeth. Like most snakes, pythons don't chase after their prey; instead, they use their senses of sight and smell to track down, and then ambush, their victims.

A hungry python will taste the air with its tongue to sense if prey is nearby. Special temperature-sensing pits on its face also tell the python if a warm-blooded animal is in the area. When its prey is close enough, the python captures it with its teeth, then swiftly wraps its body around the victim and squeezes. The python doesn't crush its prey or break its bones; rather, it tightens its coils until its prey, unable to breathe, dies of suffocation.

Incredibly, the python swallows its prey whole! It is able to unhinge its jaw and stretch its mouth wide enough to do this. Depending on the size of the python, it can swallow rodents, birds, lizards, and even large mammals such as monkeys, pigs, and antelope.

The largest python is the reticulated python of southeastern Asia and the East Indies. It can be up to thirty feet long and can weigh up to 250 pounds. Other large pythons include the African rock python, the Burmese python, and the Indian python, all of which may reach a length of twenty feet or more.

A female python can lay as many as one hundred eggs at a time! Most python mothers remain wrapped around their eggs in order to keep them warm while they develop and to help them hatch. If the temperature gets too cold, a mother python will gently squeeze its body while it is coiled around its eggs. This act, called "shivering," raises the temperature of the eggs. After the eggs hatch, however, the mother python leaves, and the newborn pythons are on their own. The mother python may not lay eggs again for another two to three years.

The average lifespan of the python is thirty-five years.

Wacky Fact:
A rock python was once discovered to have a small leopard in its stomach!

Stop!
The sight of a large cobra raised up in its warning stance is known to stop elephants in their tracks!

Baby Fact:
Cobras are the only snakes in the world that build nests for their eggs. They guard these nests fiercely until their eggs hatch.

The Cobra

Cobras are venomous snakes that can be various colors — from black or dark brown to yellowish white. The name "cobra" comes from the Portuguese term, "*cobra de capello*," which means "hooded snake," as the cobra has specialized muscles and ribs in its neck that push out the skin around its head, giving it the appearance of a hood. A cobra will raise its body high off the ground, spread its hood, and hiss loudly when threatened.

A cobra uses its forked tongue to sense the location of its prey. Its tongue moves in and out, picking up scent particles from the ground and passing them over a special smelling organ in the roof of its mouth, called the Jacobson's organ. This helps the cobra sniff out the animal that will be its meal. The cobra then kills its prey by injecting venom through its fangs. The venom is a neurotoxin that stops the prey's breathing and heartbeat. Cobras mainly eat rodents and rabbits, but will also eat birds, lizards, other snakes, and eggs.

Cobras use their venom in two different ways. Some cobras bite their prey with their fangs and in this way inject venom into them, while other cobras spit their venom at the eyes of their prey. "Spitting cobras" can spray their venom from as far as eight feet away. Their venom is not deadly, but if it comes into contact with the eyes, it can cause severe pain and temporary blindness.

There are over twenty species of the cobra. All live in Africa and Asia, and most are about six feet long. The most deadly cobra of all is the king cobra, which can reach eighteen feet in length — the longest of all venomous snakes.

The king cobra can raise up to one third of its body straight off the ground, making it taller than the average man. Although its poison is not the most potent among venomous snakes, the amount of neurotoxin it can deliver in a single bite is enough to kill twenty people! The king cobra can be found in trees, on land, and in water. It mainly eats other snakes, but it will also eat lizards, eggs, and small mammals.

Female cobras lay twelve to sixty eggs each year and usually stay close by to protect the eggs until they hatch. Once their eggs hatch, the mother cobras leave their babies to survive on their own, and the newborn cobras are able to fend for themselves. The average lifespan of a cobra is twenty years.

Did You Know?
Synthetic cobra venom is used in pain relievers and arthritis medication.

Snake / נחש

140

Hssss...
A boa constrictor can hiss so loudly that it can be heard up to 100 feet away!

Fast Fact:
The name "boa" means "large serpent" in Latin.

The Boa Constrictor

The boa constrictor is a non-poisonous constrictor snake that can be found in North, Central, and South America. Most boa constrictors are heavy snakes that are around thirteen feet long, although one was found to be as large as eighteen feet! They have very distinct and beautiful markings and colors on their bodies that vary according to their habitats. The shapes and patterns on boa constrictors include diamonds, circles, ovals, and jagged lines; while their skin colors include red, green, tan, and yellow. These distinct markings and colors help the boa constrictor blend in with its surroundings, enabling it to sneak up on its prey as well as remain hidden from its own predators.

The boa constrictor can climb trees easily and is an excellent swimmer; however, it prefers to live on dry land and makes its home mainly in hollow logs and abandoned burrows. It is called a constrictor, because after it wraps itself around its prey, it suffocates its victim by "constricting" or tightening its coils around it, until the victim is no longer able to breathe. The boa constrictor then swallows its prey whole, usually head first.

Although the boa constrictor and the python are both large constrictors, they do not belong to the same family. One difference between the two species is that the python has more teeth and one extra bone in its head. Another difference is that pythons are mostly found in the Old World (Africa, Asia, and Australia), while boas live in both the Old World and the New World (North, Central, and South America). But the biggest difference is that pythons lay eggs, while boas give birth to live young.

Most female boas give birth to thirty babies at a time, though they can give birth to as many as sixty-five babies at once. Although baby boa constrictors are not hatched from eggs, when they are born they are still surrounded by a protective membrane that they must break out of. Immediately after birth, the newborn boa constrictors are able to survive on their own. Baby boas are about two feet long when they are born, and grow continually throughout their twenty-five- to thirty-year lifespan.

Wacky Fact:
Boa constrictors like to eat bats! They catch them by hanging from tree branches and knocking the bats out of the air as they fly by.

UNIQUE TRAITS

Unique Traits

The snake is a remarkable example of how animals are created with tools so specialized, that one cannot help but marvel about it.

Take a look at the boa constrictor. Like so many snakes, the boa can swallow birds and other animals thicker than its own body because of its amazing jaw — the two halves move right apart at the hinge, and are joined only by muscle.

How does the snake manage to breathe while its mouth is engaged in the time-consuming task of swallowing a whole animal? The answer is the same specialized equipment that divers use. The boa constrictor is equipped with a snorkel-like windpipe, which it can extend to the edge of its mouth, and in this way it can breathe while swallowing its meal.

Another illustration of the snake's possessing "specialized equipment" is the viper. This poisonous snake has specially designed fangs that act as hypodermic syringes with which the snake injects poison into its victim. These fangs are so long that, when not in use, they hinge back against the roof of the viper's mouth!

The spitting cobra does not inject poison; instead, it squirts a fine stream of venom at its enemy's face, aiming for the eyes. The venom is squeezed out through holes at the tip of the snake's fangs. The spitting cobra can spit about six times before its venom supply runs out, but not to worry; this can be replaced within a day.

And then there are the baby snakes that use their teeth to saw through their shells in order to get out of their eggs all by themselves. They need no time to develop their unpleasant skills — little cobras can bite and kill as soon as they hatch from their eggs, and just one tablespoon of their dried poison could kill over 160,000 mice!

The snake is an amazing example of how each creature possesses the tools that are needed for its individual trade.

Baby snakes hatching from their egg

Baby Fact:
About 70 percent of all snakes lay eggs, while the rest give birth to live young. A baby snake is called a hatchling (a newly-hatched egg) or a neonate (a newly-born snake).

Torah Talk

And Hashem said to him, "What is in your hand?" And he said, "A staff." And He said, "Cast it to the ground," and he cast it to the ground and it became a snake. And Moshe ran away from it (Shemos 4:2-3).

The midrash relates the following incident (*Shemos Rabbah* 3): A Roman matron said to R' Yossi, "My god is greater than your G-d." He asked her why. She explained, "At the moment that your G-d revealed Himself to Moshe in the bush, Moshe covered his face [but did not move]. But when he saw the snake, which is my god, immediately, Moshe ran away from it!"

"You do not understand," R' Yossi responded. "When our G-d was revealed in the bush, there was no place where Moshe could run. Where would he run — to the heavens, the sea, or to dry land? What does it say concerning our G-d? *Behold, I fill the heavens and the earth*… With the snake, which is your god, if a person merely runs two or three steps away, he can escape and save himself, and this is why it says, *Moshe ran away from it*."

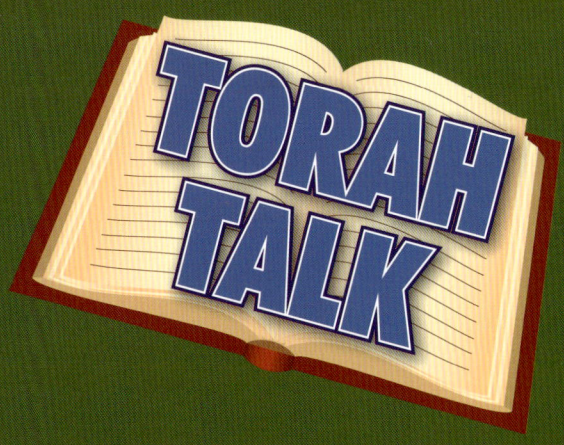

Rabbi Tzvi Elimelech Hertzberg *zt"l*, a prominent *rav* and community leader in Baltimore, Maryland in the 1940s, sees within this conversation a message about leadership. The Roman matron's god was the snake, because that was the type of leadership to which she was accustomed. Snakes will strike even without cause or personal benefit, similar to leaders of this variety: dictators, who punish their people without cause or personal benefit, but only in order to demonstrate how powerful they are.

Moshe Rabbeinu ran away from "leadership" of this nature. He wanted no part of it, for it runs completely contrary to the kindness and generosity of our forebears, the kindness demanded of us by the Torah. The Jewish path toward leadership is built upon humility, mercy, and righteousness, not the methods of a snake.

Did You Know?
Snakes are immune to their own poison.

Record Holder:
The oldest snake on record was a boa constrictor that died in 1977 at the age of forty.

INTERESTING FACTS & STATS

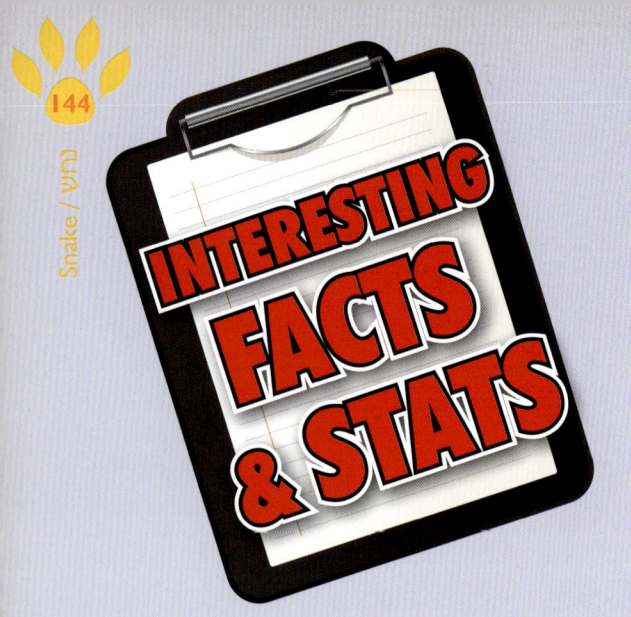

🐾 Snakes grow throughout their entire lives.

🐾 The scales of the snake act as a protective covering, allowing the snake to crawl over rough or hot surfaces like rocks, trees, and scorching desert sand.

🐾 Scales are made up of layers of cells. The outer layer of cells is dead, and it protects the living cells beneath it. A few times a year, snakes will shed a layer of their dead skin cells.

🐾 The Santa Catalina Island rattlesnake has no rattle! It's found only on Santa Catalina Island, off the coast of southern Baja California, Mexico.

🐾 After a big meal, a boa constrictor does not need to eat again for weeks.

🐾 One of the python's hunting techniques is to lie underwater in a stream, with only its head above water, and wait for a small animal to come close enough for it to attack.

🐾 Some cobra species may pretend they are dead by convulsing and then lying completely still until the danger has passed. (Most predators are not interested in dead prey; they want "fresh" prey that they've just killed by themselves.)

🐾 Ophidiophobia refers to the fear of snakes.

🐾 Cobras cannot fold their fangs down, as vipers can, so their fangs are generally shorter.

🐾 Most snakes have over two hundred teeth, which they use to hold their prey in place. Snakes are unable to chew with their teeth because their teeth are pointing backward, but they sure are able to bite with them!

Snake Trivia

1. **Which characteristic is not shared by the boa constrictor and the python?**
a. They have two lungs. b. They are not venomous. c. They are good swimmers.
d. They are mostly found in Central and South America.

2. **The anaconda, one of the world's largest snakes, is a species of…**
a. the boa b. the python c. the cobra d. the rattlesnake

3. **What do you call a group of cobras?**
a. a rhumba b. a quiver c. hoodlums d. cobrettes

4. **How often does a rattlesnake usually give birth?**
a. once a month b. once a year c. twice a year d. once every two to three years

5. **Which statement is false?**
a. Snakes are reptiles. b. Snakes cannot blink. c. Snakes have ears.
d. Most snakes are not venomous.

ANSWERS: 1. d 2. a 3. b 4. d 5. c

Animal Crackers:
Q. What did the snake wear to his wedding?
A. A boa tie.

WEASEL סמור

The Weasel

The weasel is a small-sized, meat-eating mammal that lives in various habitats, including grasslands, sand dunes, forests, farmlands, and even small cities. The weasel is the most widespread carnivore in the Western Hemisphere, and is also found in many parts of the Northern Hemisphere.

A weasel has a small, flat, triangular head with short, rounded ears, and long whiskers. It has brown fur on the upper part of its body and white to yellowish fur on its undersides.

In cold regions, its fur turns white in the winter. A weasel can weigh between three to seven ounces, and, depending on the species, can be as little as five inches or as long as twenty-four inches (including its tail).

Weasels communicate with each other through various methods. One way is by its odor. Weasels produce musk, which is a thick, oily, yellowish fluid with a very distinct scent. The weasel uses its musk to mark the borders of its territory. Should another weasel enter its territory, the odor sends a message to the intruder that it is trespassing and should keep out — or else! Weasels also use body language and vocal sounds to communicate with each other.

There are numerous species of the weasel. The three most common weasels are the long-tailed weasel, the short-tailed weasel, and the least weasel which is also referred to as the common weasel. The least weasel is the smallest of all weasels, but it is the most wide-spread. Unlike the other weasels, the least weasel does not have a black tip on the end of its tail.

The long-tailed weasel is the largest of the three species, due to its tail which comprises over forty percent of its body length. Long-tailed weasels in the southwest of the United States have a white mask.

The short-tailed weasel, which is also called an ermine, is smaller than the long-tailed weasel, but larger than the least weasel. Like other weasels, short-tailed weasels have a brown coat in the summer and a white coat in the winter (in cold regions). They measure between seven and thirteen inches in length.

Though weasels climb trees to catch prey, to escape predators, or even just to rest, they are most at home on and under the ground.

Did You Know?

Weasels are nearsighted. This means that they are able to see objects that are close to them but are unable to see distant objects clearly.

THE CIRCLE OF LIFE

The Circle of Life

The female weasel gives birth in the spring. Most litters consist of four to eight baby weasels, called *kits* or kittens. The kits are incredibly small and only weigh one tenth of an ounce or less! The kits are born blind, deaf, and with a little covering of fur.

The kits grow very fast. The mother weasel brings her kits meat to eat after they are weaned, which is usually when they are about one month old. At five weeks of age, the kits begin to join their mother on hunts for food. By the time the kits are eight weeks old, they are able to hunt and kill on their own.

When the kits are between nine and twelve weeks of age, the family of weasels splits up as the mother sends her kits off in order for them to find homes of their own. By the time the weasel reaches four months of age, it is fully grown.

Although the weasel is agile and fast, its small size makes it very vulnerable to a host of predators that share its environment. Common land predators of the weasel include snakes and foxes. Weasels also have to be on the alert for aerial predators such as eagles, hawks, and owls.

Female weasels live three to four years in the wild. The male weasels, however, usually survive only about one year in the wild. The males are more prone to attack by predators since they need to travel in open areas in search of a mate.

What's for Supper?

Weasels prefer to eat only the animals that they kill on their own, as opposed to scavengers that eat any meat they can find. Weasels are very active, and therefore need to eat at least one third of their body weight every day in order to have sufficient energy to survive.

The weasel's diet consists of a large assortment of animals, including mice, frogs, chipmunks, squirrels, gophers, and lemmings. Weasels will even eat larger animals such as rabbits and ducks, and occasionally they'll eat smaller creatures such as insects and worms.

Weasels also like eggs and will invade the nests of their prey to get to them. Weasels are able to eat large eggs by biting a hole in one or both ends of the egg and sucking out the contents, leaving the shell behind.

Baby Fact:
The female weasel takes care of its kits all on her own. The father does not take part in raising its young.

Unique Traits

The weasel is the smallest carnivorous mammal in the world, and yet, despite its tiny size, it is an extremely vicious and ruthless hunter. It is a solitary animal that spends most of its life preying on small animals during both the day and night.

One of the weasel's unique hunting techniques is that it can enter the burrows of its prey — something that most predators are unable to do. Its long, slender body is perfectly suited for following its prey into their tunnels and homes. The weasel has a small and narrow head, short legs, and a flexible spine which enables it to get into small and narrow openings. Once the weasel can smell its prey, it will not let up until it tracks it down, often following the trail of its prey right into their burrow or den.

Like all professional hunters, the weasel is equipped with an array of lethal weapons. It has sharp, pointed canine teeth, which are used for biting and tearing flesh. A weasel's teeth can kill an animal more than double its size! The weasel's arsenal also includes ten toes that are tipped with small, non-retractable claws.

The weasel is a strong and ferocious fighter, and is able to move extremely fast. It is fearless and has been known to attack animals much larger and stronger than itself, such as bears and porcupines. If a weasel feels threatened, it will even attack humans.

The weasel can also be a stealthy hunter. It is able to sneak up on its prey without making a sound before launching an attack. It sometimes uses camouflage to spring up on its prey. During the summer months, the weasel is light brown, which matches the fields of its surroundings. During the winter months, the weasel's coat turns white, to match the snow. Only the tip of its tail remains black throughout the year.

This black tip, however, helps the weasel escape from birds of prey. A bird of prey will spot the black tip of the tail and mistake it for the weasel's head. The bird will then dive toward the weasel's "head" to attack, unknowingly focusing its attack on the non-vulnerable tail instead. When the bird misses like this, the weasel is able to run forward and away from the bird, unharmed.

What's That Smell? The odor of a weasel is as bad as a skunk's smell!

Wacky Fact: The weasel will sometimes catch its prey by dancing! The "Weasel War Dance" is a play dance put on by weasels to confuse their prey. The weasel bounces all about, flipping over and hopping sideways. Once its prey becomes mesmerized, the weasel attacks.

Torah Talk

There is an allusion in the Gemara (*Taanis* 8a) to a fascinating story about a weasel and a well. The story begins: *Rav Ami said, "Come and learn the greatness of men and faith from the story of the weasel and the well."*

On her way home one day, a teenage girl fell into a well. She called for help, and her cries attracted the attention of a young man who happened to pass by at the time. He offered to help her if she promised to marry him.

The girl agreed to his condition, and he pulled her up. As there were no witnesses around to testify to the promise the two had made, the young man suggested that the well and a weasel that was seen near it should be invoked as witnesses. This done, the couple parted ways.

The girl kept her promise and did not marry anyone else, waiting for the young man to return to marry her; however, the young man forgot about the incident and moved to another town. Shortly thereafter, he married another girl, who bore him a son. At the age of three months, his son was bitten by a weasel and died. Later, a second boy was born to the couple, and he fell into a well and drowned. The wife, becoming alarmed by the peculiar nature of these accidents, asked her husband for some explanation. He then remembered the oath that he'd once made and told the story to his wife. After hearing the entire incident, the wife demanded a divorce and obtained it.

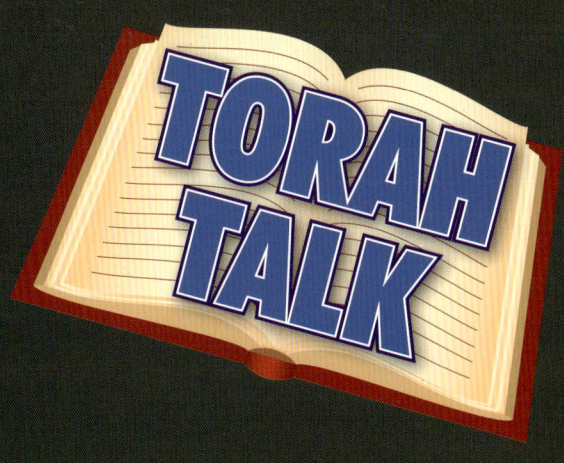

During all that time, the young man's former fiancée was waiting for his appearance, discouraging all her suitors by faking epileptic fits. When the young man, after a long search, finally found her, she did not recognize him and tried to deter him in the same way. Then he mentioned to her about the weasel and the well — and at once she accepted him. They got married and lived happily together, and were blessed with children (compare Rashi and *Tosafos*).

Fast Fact:
Weasels are very good swimmers.

INTERESTING FACTS & STATS

- Weasels are very quick and are noted for their lightning-speed movements.
- A male weasel can be called a dog, buck, hob, or jack.
- A female weasel can be called a doe or jill.
- Weasels run by taking a series of leaps, with their backs humped at each movement and their tails trailing backward.
- Weasels often stand on their hind legs in order to search for food and to look out for predators.
- Male weasels are much larger than female weasels. They weigh nearly twice as much and their body length is a quarter longer.
- The weasel uses its long whiskers as a guide while traveling in the dark.
- Since ancient times, the fur of weasels has been used for making heavy winter jackets.
- In some societies, the fur of a weasel was an indication of royalty.
- Weasels' nests are made of grass and leaves and can sometimes be found in tree stumps.
- Often, weasels do not need to make their own nest, as they use the nest, tunnel, or burrow of one of the animals they have eaten.
- Weasels have been introduced to countries where they are not naturally found, due to their versatile and dominant nature as a form of a natural pest and rodent control.

Weasel Trivia

1. Which one of these statements is false?
a. Weasels are good swimmers. b. Weasels are good climbers. c. Weasels have poor night vision. d. Weasels have a great sense of smell.

2. By what age are baby weasels able to hunt on their own?
a. two weeks old b. four weeks old c. eight weeks old d. twelve weeks old

3. The weasel mostly attacks its prey…
a. on level ground b. underground c. up in trees d. in the water

4. Which weasel is usually the smallest in size?
a. the common weasel b. the short-tailed weasel c. the long-tailed weasel d. the ermine

5. What do you call a group of weasels?
a. a problem b. a herd c. a mob d. a gang

ANSWERS: 1. c 2. c 3. b 4. a 5. d

Animal Crackers:
Q. Where did the long-tailed weasel go when he cut his tail?
A. To a re-tail store.

WOLF　　　זאב

The Wolf

Wolves live in family groups called packs. A wolf pack generally consists of an adult male, an adult female (a mother and father), and their young. The average size of a pack is eight to twelve wolves, but packs can contain as many forty wolves. The father is the leader of the pack and is called the alpha male, while the mother is referred to as the alpha female. Sometimes unrelated wolves join a pack, but the parents are usually the leaders, since they tend to be the strongest and smartest in the group.

Wolves are intelligent creatures and very skillful hunters. They have excellent senses of hearing and vision, and are able to pick up on the slightest movements. Their eyes are set at the front of their heads, so they are able to spot their prey with both eyes at once. They also have very good night vision. In addition, wolves have an amazing sense of smell. They can pick up the scent of a deer from over one mile away!

Wolves have strong stamina and are able to run long distances at a fast pace. A wolf can run for twelve miles at 15-30 mph. Over a short distance, it can sprint as fast as 40 mph.

Wolves usually hunt in a pack. This makes them more efficient hunters. Pack members work well together and use various techniques to capture their prey. Some wolves may distract a mother animal while the others sneak up on her young. Other times, some wolves will go ahead of the pack and hide. They capture their prey when the rest of the pack drives it toward them. Wolves usually choose a weak or old animal as their prey, since it is easier to catch than a young, healthy one.

There are two different wolf species in North America: the gray wolf and the red wolf. The coat of the gray wolf is usually light gray, sprinkled with black, although it may be pure white in the Arctic. Gray wolves are members of the dog family, which includes dogs, foxes, jackals, and coyotes. They are the largest of all wild dogs. The red wolf is a smaller and a more slender cousin of the gray wolf. It is gray-black, with a reddish cast that gives it the color for which it is named.

Did You Know?

Wolves can hear a sound as far as six miles away in the forest, and ten miles away in the open country.

THE CIRCLE OF LIFE

The Circle of Life

The average litter size of a female wolf is four to seven pups; however, it can be as large as fourteen pups. A newborn pup only weighs one pound at birth and is born blind and deaf. The mother wolf does not leave her pups while they are helpless, except to eat the food that her pack members bring her.

At around two weeks of age, the pups are able to see, and at around three weeks they are able to hear. Between three and five weeks of age, the pups begin to venture outside their den and eat meat. Both parents, as well as the other adult family members in the pack, take part in raising the pups.

A "pup-sitter" watches over the pups while the mother goes out hunting. The pup-sitter is usually a "lower-class" member of the pack. It may be a sibling from the previous year's litter.

Wolf pups play a lot as they grow up. While they're having fun, they're also sharpening their survival skills. By playing, they learn how to communicate and get along with each other. Their games provide practice for hunting techniques. When they are around three months old, wolf pups begin to join the adults on hunting trips, and start to actively hunt at seven to eight months of age. The pups stay with their parents for up to two years.

The average lifespan of the wolf in the wild is ten years.

What's for Supper?

Wolves are carnivorous animals. They generally hunt large animals, but will also hunt smaller ones if necessary. Wolves hunt together in packs and work together as a team in order to catch and kill their prey. Their prey includes moose, caribou, elk, bison, and mountain goats. At times they will eat hares, beavers, birds, insects, and berries.

When wolves find food, they eat as much as they can, because they may go for several days before eating again. Wolves can kill their prey effectively because they have extremely powerful jaws, which can bite through even the toughest of animal hides. They are even able to crack hard bones in just a few bites.

Baby Fact:
Baby wolves can utter their first howls at four weeks of age.

UNIQUE TRAITS

Unique Traits

Wolves possess a unique set of communication skills. Their communication techniques are vital for the pack's survival, since the wolves need to work together to hunt for food, raise their young, and protect their territory. They are able to communicate using various methods that include body language, howling, and sounds.

Body Language

Wolf packs follow the lead of the head male and head female, called the alphas. The alphas are responsible for organizing the pack to hunt as a group. Good communication among pack members helps keep order and allows the pack to work together efficiently. When pack members greet their leader, they use body language to show him honor. They crouch down with their ears flattened and tails low. This is their way of stating, "You are the leader." When a wolf feels confident, however, it will approach another wolf with its head and tail held high and ears raised.

Wolves also use facial expressions to convey their feelings. They will show off their teeth and growl if they are angry. If a wolf's ears are flattened back against its face, it is an indication that it is being cautious or that it is frightened.

Howling

One of the best-known ways that wolves communicate is by howling. Their howling can be heard from up to ten miles away, and it is done for many reasons. One reason is for wolves to claim their territory, as their howls warn other wolves to stay away. Wolves will howl to gather the pack before and after a hunt for prey. Howling also helps pack members keep in touch when they get separated, and it enables them to reunite.

Wolves have a variety of pitched howling sounds to indicate their actions. Howls used for calling pack mates to hunt are long, smooth sounds. While they are chasing their prey, they let out a higher-pitched howl. When closing in on their prey, they emit a combination of a short bark and a howl.

Other Sounds

In addition to howling, wolves also use other sounds to communicate. They whimper, whine, growl, and bark. Whining is associated with situations of nervousness, curiosity, and friendliness such as when greeting, feeding young, or playing. Growling is usually made during food challenges, such as when wolves fight over food among themselves, whereas barking is generally heard when wolves are startled.

Myth Buster
Wolves do not howl at the moon.

Wacky Fact:
If threatened, a wolf will empty out the contents of its stomach in order to make its body lighter. This allows it to run faster and escape unharmed.

Torah Talk

...recalling the sins of the fathers on the children... (*Bamidbar* 14:18).

There was once a hungry wolf that was about to eat a fox. The fox said to the wolf, "What good is my skinny body going to do you? Look over there at that nice, juicy human with his body so full with meat and fat! Better to eat him. He's much more of a meal than me!"

The wolf answered, "It is forbidden for us to eat humans. In the end I will get punished for it."

"Don't worry about it!" replied the sly fox. "You won't get punished. Only your children will get punished, as it says, *The fathers ate sour grapes, but it is the teeth of the sons that are set on edge*" (*Yirmiyah* 31:28).

Instead of being highly surprised and somewhat cautious of a fox that could quote Tanach by heart, the wolf instead fell for this line of reasoning and, licking his lips, he made off toward the man. No sooner had he taken a few paces, then he fell into a carefully concealed trap and found himself plunging into a deep pit.

The wolf wailed up at the fox, "You lied to me! You said that only my children would be punished and not me!"

The fox answered, "Fool that you are! This punishment isn't because of what you did. It's because of what your father did before you."

"How can that be possible?" the wolf cried bitterly. "I have to suffer for the sins of my father?"

The fox replied, "You were prepared to eat a man, though you knew your children would suffer for it in the future, so why should you complain for getting punished now for what your father did before you?"

From this parable, we can perhaps understand the difficult verse quoted above from *Parshas Shelach*. For where indeed is the justice in Hashem's *recalling the iniquity of the fathers on the children*? Even if the children perpetuate the misdeeds of their fathers, shouldn't they be punished for their own sins only, and not for those of their fathers? Why should they be responsible for their fathers' behavior? However, if the children understand that their misdeeds can cause their own children to suffer, they can have no complaint when they themselves must pay for the sins of their fathers.

Record Holder:

In the winter of 2010-11, a "super pack" of 400 wolves reportedly terrorized the Russian town of Verkhoyansk. This is by far the largest pack of wolves ever recorded.

INTERESTING FACTS & STATS

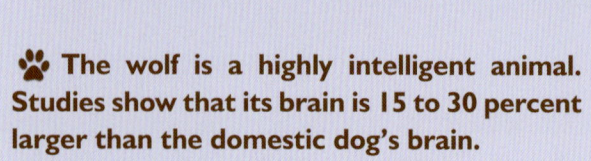

🐾 The wolf is a highly intelligent animal. Studies show that its brain is 15 to 30 percent larger than the domestic dog's brain.

🐾 Packs hunt in territories of up to six hundred square miles.

🐾 A pack may cover distances from thirty to one hundred and twenty-five miles in a day.

🐾 There are two divisions in a wolf pack, one for females and one for males.

🐾 A wolf den is often near a river or lake, so the mother wolf does not have to go far to get water.

🐾 A wolf often eats about twenty pounds of food at one meal.

🐾 Surprisingly, wolves have a low hunting-success rate.

🐾 When adult wolves return from a hunt, the pups lick the adults' mouths to encourage them to regurgitate undigested meat, which the pups then eat.

🐾 There is no record of a healthy wolf ever eating a human being in the United States.

🐾 Arctic wolves are not born white. Newborn pups are a gray-brown color. Since they are usually born after the snow melts, this color is more conducive to helping camouflage, or hide, them during their first summer.

🐾 In the winter, the wolf uses its tail to keep its face warm.

Wolf Trivia

1. What is the lowest-ranking member of the wolf pack called?
a. omega b. alpha c. delta d. pup

2. How many teeth does a wolf have?
a. 32 b. 42 c. 102 d. A wolf does not have teeth.

3. The gray wolf is also referred to as…
a. a red wolf b. a werewolf c. a wolverine d. a timber wolf

4. Which statement is false?
a. Wolves have a good sense of smell. b. Wolves have a good sense of hearing.
c. Wolves have four toes on each paw. d. Wolves are good swimmers.

5. Which member of the pack is responsible to take care of the young pups?
a. the mother b. the father c. the adult siblings d. answers a, b, and c

ANSWERS: 1. a 2. b 3. d 4. c 5. d

Animal Crackers:
Q. What did one wolf say to another?
A: Howl's it goin'?

Sources:

Apes
Torah Talk – Adapted from *Rabbi Kornfeld's Weekly Parasha Page, Parashat Va'era*, 5755 – Vilna Gaon: *Adnei Hasadeh*; and from *Ohr Somayach International*, with permission given by Editor, Rabbi Moshe Newman.

Bears
Torah Talk – Adapted from *Rabbi Mordecai Kornfeld's Weekly Parasha Page. Parashat Va'era, 5755* – Maharil Diskin: Bears without a forest.

Beaver
Unique Traits - Adapted from *Our Wondrous World*, by Rabbi Avrohom Katz, with permission of the copyright holders, *Artscroll / Mesorah Publications, Ltd.*
Torah Talk – Adapted from an article written by Rabbi Raymond Beyda, with permission from *Project Genesis – Torah.org* (where a version of it first appeared).

Camel
Unique Traits – Adapted from *Our Amazing World*, by Rabbi Avrohom Katz, with permission of the copyright holders, *Artscroll / Mesorah Publications, Ltd.*
Torah Talk – Adapted from *The Animal Kingdom in Jewish Thought*, by Shlomo Pesach Toperoff. Published by Jason Aronson Publishers, Inc. Jason Aronson is an imprint of Rowman & Littlefield Publishers, Inc.

Cow
Circle of Life and Unique Traits – Adapted from *Our Amazing World*, by Rabbi Avrohom Katz, with permission of the copyright holders, *Artscroll / Mesorah Publications, Ltd.*
Torah Talk – Adapted from an article written by Rabbi Pinchas Winston, with permission from *Project Genesis - Torah.org* (where a version of it first appeared).

Elephant
Unique Traits – Adapted from *Designer World*, by Rabbi Avrohom Katz.

Fox
Torah Talk – Adapted from *The Animal Kingdom in Jewish Thought*, by Shlomo Pesach Toperoff. Published by Jason Aronson Publishers, Inc. Jason Aronson is an imprint of Rowman & Littlefield Publishers, Inc.

Gazelle
Torah Talk – Adapted from *The Animal Kingdom in Jewish Thought*, by Shlomo Pesach Toperoff. Published by Jason Aronson Publishers, Inc. Jason Aronson is an imprint of Rowman & Littlefield Publishers, Inc.

Giraffe
Unique Traits – Adapted from *Our Amazing World*, by Rabbi Avrohom Katz, with permission of the copyright holders, *Artscroll / Mesorah Publications, Ltd.*
Torah Talk – *Ohr Somayach Institutions* / www.ohr.edu

Goat
Torah Talk – Adapted from an article written by Rabbi Yosef Kalatsky, with permission from *Project Genesis - Torah.org* (where a version of it first appeared).

Hippopotamus
Torah Talk – Adapted from an article written by Rabbi Tzvi Gluckin; and from an article by the *Chicago Rabbinical Council*, the Association of Kashrus Organizations.

Horses
Unique Traits – Adapted from *The Animal Kingdom in Jewish Thought*, by Shlomo Pesach Toperoff. Published by Jason Aronson Publishers, Inc. Jason Aronson is an imprint of Rowman & Littlefield Publishers, Inc.
Torah Talk – Adapted from *Project Genesis – Torah.org* (where a version of it first appeared).

Kangaroo
Unique Traits – Adapted from *Our Amazing World*, by Rabbi Avrohom Katz, with permission of the copyright holders, *Artscroll / Mesorah Publications, Ltd.*
Torah Talk – Adapted from *Rabbi Kornfeld's Weekly Parasha Page, Parashat Va'era, 5755* – The Rogatchover: A source of food.

Leopard
Unique Traits – Adapted from *Noah's Ark*. To learn more about Judaism and animals, visit www.chabad.org/NoahsArk.
Torah Talk – Adapted from Nathan Elberg's article in *The Jewish Press*, 11-22-06.

Lion
Torah Talk – Adapted from *The Animal Kingdom in Jewish Thought*, by Shlomo Pesach Toperoff. Published by Jason Aronson Publishers, Inc. Jason Aronson is an imprint of Rowman & Littlefield Publishers, Inc.

Monkeys
Torah Talk – Sources taken from *YU Torah*, from an article written by Rabbi Michael Taubes, titled 'Visiting the Zoo.'

Porcupine
Unique Traits – Adapted from *Our Amazing World*, by Rabbi Avrohom Katz, with permission of the copyright holders, *Artscroll / Mesorah Publications, Ltd.*

Snakes
Unique Traits – Adapted from an article written by Rabbi Yaakov Menken, with permission from *Project Genesis - Torah.org* (where a version of it first appeared).
Torah Talk – Adapted from *Our Amazing World*, by Rabbi Avrohom Katz, with permission of the copyright holders, *Artscroll / Mesorah Publications, Ltd.*

Weasel
Torah Talk – Adapted from *The Animal Kingdom in Jewish Thought*, by Shlomo Pesach Toperoff. Published by Jason Aronson Publishers, Inc. Jason Aronson is an imprint of Rowman & Littlefield Publishers, Inc.

Wolf
Torah Talk – based on *Pardes Yosef* in *Mayanah Shel Torah* – page 71 (*Bamidbar* 14:18). Adapted from an article written for *Ohr Somayach Institutions* / www.ohr.edu

* **Secular information was obtained in part by the online animal encyclopedia – www.a-z-animals.com.**

** **Except for quoted factual data, the contents of secular websites named as sources are not endorsed by the Author or the Publisher.**

Photo Credits

Apes
Page 7: Stockbyte/Thinkstock, iStockphoto/Thinkstock, Hermera / Thinkstock, Zoonar/Thinkstock
Page 8: Tom Brakefield/Stockbyte/Thinkstock
Page 9: Jupiter Images/Photos.com/Thinkstock
Page 10: iStockphoto/Thinkstock
Page 11: iStockphoto/Thinkstock
Page 12: iStockphoto/Thinkstock
Page 13: iStockphoto/Thinkstock
Page 14: iStockphoto/Thinkstock
Page 15: Zoonar/Thinkstock
Page 16: iStockphoto/Thinkstock, Hermera/Thinkstock, Anup Shah/Digital Vision/Thinkstock, iStockphoto/Thinkstock
Page 17: David Davis/Bigstock,
Page 18: iStockphoto/Thinkstock, iStockphoto/Thinkstock

Bears
Page 19: iStockphoto/Thinkstock, iStockphoto/Thinkstock, Hemera/Thinkstock, iStockphoto/Thinkstock
Page 20: iStockphoto/Thinkstock
Page 21: iStockphoto/Thinkstock
Page 22: Glen Gafney/Bigstock
Page 23: iStockphoto/Thinkstock
Page 24: iStockphoto/Thinkstock
Page 25: iStockphoto/Thinkstock
Page 26: iStockphoto/Thinkstock
Page 27: iStockphoto/Thinkstock
Page 28: iStockphoto/Thinkstock, iStockphoto/Thinkstock, Hemera/Thinkstock, iStockphoto/Thinkstock
Page 29: iStockphoto/Thinkstock
Page 30: iStockphoto/Thinkstock, iStockphoto/Thinkstock

Beaver
Page 31: iStockphoto/Thinkstock
Page 32: Ammit/Bigstock
Page 33: Ingram Publishing/Thinkstock
Page 34: Hemera/Thinkstock, Gvision/Dreamstime
Page 35: Kwiktar/Dreamstime
Page 36: Ammit/Bigstock, Hemera/Thinkstock

Camel
Page 37: Hemera/Thinkstock
Page 38: Galyna Andrushko/Bigstock
Page 39: iStockphoto/Thinkstock, iStockphoto/Thinkstock
Page 40: iStockphoto/Thinkstock
Page 41: Zoonar/Thinkstock
Page 42: iStockphoto/Thinkstock, Regien Paassen/Bigstock

Cow
Page 43: iStockphoto/Thinkstock
Page 44: Hemera/Thinkstock
Page 45: Zoonar/Thinkstock, Digital Vision/Thinkstock
Page 46: Hemera/Thinkstock
Page 47: John Foxx/Stockbyte/Thinkstock
Page 48: Hemera/Thinkstock, iStockphoto/Thinkstock

Elephant
Page 49: iStockphoto/Thinkstock
Page 50: Hemera/Thinkstock, iStockphoto/Thinkstock
Page 51: iStockphoto/Thinkstock
Page 52: iStockphoto/Thinkstock
Page 53: iStockphoto/Thinkstock
Page 54: iStockphoto/Thinkstock, iStockphoto/Thinkstock

Fox
Page 55: iStockphoto/Thinkstock
Page 56: Jupiter Images/Photos.com/Thinkstock
Page 57: iStockphoto/Thinkstock, iStockphoto/Thinkstock
Page 58: iStockphoto/Thinkstock
Page 59: Tom Brakefield/Stockbyte/Thinkstock
Page 60: iStockphoto/Thinkstock, Altrendo Nature/Stockbyte/Thinkstock

Gazelle
Page 61: iStockphoto/Thinkstock
Page 62: Digital Vision/Thinkstock
Page 63: iStockphoto/Thinkstock, Getty Images/Jupiter Images/Photos.com/Thinkstock
Page 64: iStockphoto/Thinkstock
Page 65: iStockphoto/Thinkstock
Page 66: iStockphoto/Thinkstock, Getty Images/AbleStock.com/Thinkstock

Giraffe
Page 67: iStockphoto/Thinkstock
Page 68: Art Wolfe/Lifesize/Thinkstock
Page 69: iStockphoto/Thinkstock, Zoonar/Thinkstock
Page 70: Zoonar/Thinkstock
Page 71: Anup Shah/Photodisc/Thinkstock
Page 72: iStockphoto/Thinkstock, iStockphoto/Thinkstock

Goat
Page 73: iStockphoto/Thinkstock
Page 74: iStockphoto/Thinkstock
Page 75: Getty Images/Jupiter Images/Photos.com/Thinkstock, Design Pics/Thinkstock
Page 76: Ints Tomsons/Bigstock
Page 77: Miruslav Hiadik/Bigstock
Page 78: iStockphoto/Thinkstock, iStockphoto/Thinkstock

Hippo
Page 79: iStockphoto/Thinkstock
Page 80: iStockphoto/Thinkstock
Page 81: iStockphoto/Thinkstock, iStockphoto/Thinkstock
Page 82: iStockphoto/Thinkstock
Page 83: Hemera/Thinkstock
Page 84: Getty Images/Photos.com/Thinkstock, Tom Brakefield/Stockbyte/Thinkstock

Horses
Page 85: iStockphoto/Thinkstock, iStockphoto/Thinkstock, iStockphoto/Thinkstock, iStockphoto/Thinkstock
Page 86: iStockphoto/Thinkstock
Page 87: iStockphoto/Thinkstock
Page 88: iStockphoto/Thinkstock
Page 89: iStockphoto/Thinkstock
Page 90: Carolyne Pehora/Bigstock
Page 91: iStockphoto/Thinkstock
Page 92: iStockphoto/Thinkstock
Page 93: iStockphoto/Thinkstock
Page 94: iStockphoto/Thinkstock
Page 95: iStockphoto/Thinkstock
Page 96: iStockphoto/Thinkstock, iStockphoto/Thinkstock

Kangaroo
Page 97: iStockphoto/Thinkstock
Page 98: iStockphoto/Thinkstock
Page 99: iStockphoto/Thinkstock, iStockphoto/Thinkstock
Page 100: iStockphoto/Thinkstock, Hemera/Thinkstock
Page 101: iStockphoto/Thinkstock
Page 102: Hemera/Thinkstock

Leopard
Page 103: iStockphoto/Thinkstock
Page 104: Design Pics/Thinkstock
Page 105: Anup Shah/Digital Vision/Thinkstock
Page 106: Tom Brakefield/Stockbyte/Thinkstock
Page 107: Tome Brakefield/Stockbyte/Thinkstock
Page 108: Ryan McVay/Digital Vision/Thinkstock, John Foxx/Stockbyte/Thinkstock

Lion
Page 109: iStockphoto/Thinkstock
Page 110: iStockphoto/Thinkstock
Page 111: iStockphoto/Thinkstock
Page 112: Hemera/Thinkstock
Page 113: Tom Brakefield/Stockbyte/Thinkstock
Page 114: iStockphoto/Thinkstock, iStockphoto/Thinkstock

Monkeys
Page 115: iStockphoto/Thinkstock, iStockphoto/Thinkstock, iStockphoto/Thinkstock, iStockphoto/Thinkstock
Page 116: Anup Shah/Digital Vision/Thinkstock
Page 117: iStockphoto/Thinkstock
Page 118: iStockphoto/Thinkstock
Page 119: Anup Shah/Photo Disc/Thinkstock
Page 120: iStockphoto/Thinkstock
Page 121: iStockphoto/Thinkstock
Page 122: Anup Shah/Digital Vision/Thinkstock
Page 123: iStockphoto/Thinkstock
Page 124: iStockphoto/Thinkstock , Ryan McVay/Digital Vision/Thinkstock
Page 125: iStockphoto/Thinkstock
Page 126: iStockphoto/Thinkstock, Ryan McVay/Digital Vision/Thinkstock

Porcupine
Page 127: Tom Brakefield/Stockbyte/Thinkstock
Page 128: Getty Images/Jupiter Images/Photos.com/Thinkstock
Page 129: iStockphoto/Thinkstock, iStockphoto/Thinkstock
Page 130: Digital Vision/Thinkstock
Page 131: Getty Images/Jupiter Images/Photos.com/Thinkstock
Page 132: Thomas Wodruff/Bigstock, Tom Brakefield/Stockbyte/Thinkstock

Snakes
Page 133: iStockphoto/Thinkstock, Design Pics/Thinkstock, iStockphoto/Thinkstock, Hemera/Thinkstock
Page 134: iStockphoto/Thinkstock
Page 135: Altrendo Nature/Stockbyte/Thinkstock
Page 136: iStockphoto/Thinkstock
Page 137: iStockphoto/Thinkstock
Page 138: iStockphoto/Thinkstock
Page 139: iStockphoto/Thinkstock
Page 140: Getty Images/Jupiter Images/Liquid Library/Thinsktock
Page 141: iStockphoto/Thinkstock
Page 142: Getty Images/Jupiter Images/Photos.com/Thinkstock
Page 143: iStockphoto/Thinkstock
Page 144: iStockphoto/Thinkstock, iStockphoto/Thinkstock

Weasel
Page 145: Altrendo Nature/Stockbyte/Thinkstock
Page 146: iStockphoto/Thinkstock
Page 147: iStockphoto/Thinkstock, iStockphoto/Thinkstock
Page 148: Thomas Woodruff/Bigstock
Page 149: Tom Brakefield/Stockbyte/Thinkstock
Page 150: iStockphoto/Thinkstock, Hemera/Thinkstock

Wolf
Page 151: iStockphoto/Thinkstock
Page 152: Tom Brakefield/Stockbyte/Thinkstock
Page 153: iStockphoto/Thinkstock, iStockphoto/Thinkstock
Page 154: iStockphoto/Thinkstock
Page 155: Comstock/Thinkstock
Page 156: Design Pics/ iStockphoto/Thinkstock